LAND OF PLENTY

LAND OF PLENTY

A Journey Through the Fields
& Foods of Modern Britain

CHARLIE PYE-SMITH

First published 2017 by
Elliott and Thompson Limited
27 John Street
London WC1N 2BX
www.eandtbooks.com

ISBN: 978-1-78396-305-8

Page xvi: R. S. Thomas, 'A Peasant', 1942. From R. S. Thomas, *Collected
Poems 1945–1990*, W&N, 2000. Page 74: Sylvia Plath, 'Wuthering Heights',
1961. From *The Selected Poems of Sylvia Plath*, Faber & Faber, 2003.

9 8 7 6 5 4 3 2 1

A catalogue record for this book is available from the British Library.

Typesetting: Marie Doherty
Printed in the UK by TJ International Ltd

For Sandie, Sophie and Chris

Contents

Introduction

I was brought up in a small valley on the outskirts of Huddersfield, a wool town in the West Riding of Yorkshire. When I reached the age of six or seven, in the late 1950s, I used to wander where I pleased among the fields and farms. Nobody seemed to mind. I watched, with a mixture of horror and fascination, the goings-on at the pig farm behind our back garden owned by Mr Codd. I still shudder at the memory of him pulling piglets out of the shell of an abandoned post office van before slicing off their testicles with a razor and feeding them to his dogs. More pleasurable were the frequent jaunts down to the farm in the valley bottom owned by Mr and Mrs Sheard, who delivered our daily milk and eggs. They were kindly folk and would let me and my friends scramble around the farm, peer into the cow byre and look for sticklebacks in the stream below their farmhouse.

My early enthusiasm for agricultural matters was kindled on farmhouse summer holidays, always taken at harvest time, first in North Yorkshire, later in West Wales and Ross and Cromarty. From the moment we arrived to the moment we left I would busy myself around the farms – collecting eggs, feeding pigs, watching cows calve, rounding up sheep, stacking straw. This, I decided, was the life for me and I applied for a place at Newcastle University to study agriculture.

Before I went there, I spent a year on a farm in the Yorkshire Dales, near the market town of Masham. It was the best of times, it was the worst of times, though not in that order. I began work on the Clifton Estate just after the cereal harvest and spent the first five or six weeks chisel-ploughing stubble fields, alone with my thoughts, the smell of diesel and the autumnal weather. In November I was transferred to the dairy, where I was entrusted with mostly mundane tasks: gathering the cows at 5 o'clock in the morning and bringing them in for milking; forking silage, sometimes in sleet and snow; feeding calves; scraping slurry and washing out the parlour. From time to time I was allowed to milk the cows, but the more temperamental among them sensed my lack of experience and lashed out with their hooves. I had a miserable winter.

But all this changed with the coming of spring. I was transferred back to the arable side of the farm and I recall the summer of 1970 with nostalgic clarity. My memories are not so much of the farm work – although there were many activities I loved, especially at hay-time and harvest – but of the chatter with my fellow farmworkers, one of whom is still a great friend. In those days there was the daily ritual of 'allowance' or 'bait' time when we ate our mid-morning sandwiches. In busy periods this never amounted to more than a ten-minute break, but during the quieter times of year it could be spun out for thirty or forty minutes before the foreman, Percy Vayro, would declare, hurriedly pulling a fob watch from his waistcoat, that it was time to get back to work. The best way to get a long bait time was to encourage Percy or one of the older farmworkers – some were born just after the First World War – to talk about days gone by. You only had to mention horses – 't'owd yasses', as Percy would call them – to get the most intricate explanations of what it was like ploughing with horses, what you had to do to make them pee at the end of a row, and other interesting quirks of equine behaviour.

I even enjoyed some of the drearier tasks, such as hoeing turnips in the company of Bill Thorburn, who had just been released from Dartmoor Prison, where he had served time for manslaughter, and picking wild oats in the company of Rabbit Close and a gang of casual labourers, including a young man who used to turn up with a ferret squirming in his coat pocket. It is over forty years since I saw most of these characters and many are now dead, but I can remember their accents and laughter and some of the conversations we had, just as I can instantly recall, if I close my eyes, the acrid odour of the silage pit, the sweet aromas that rose from newly mown hay meadows, the clack-clack of the grass cutter, the milky smell of a warm cow's udder, the foul stench of a calf with scour, the dusty nose-clogging air in the corn-drying shed.

By the end of the year I knew that I would make an incompetent farmer, so when I got to university I changed subjects – to botany and zoology – and pursued a very different career. Nevertheless, farming remains the first love of my life and I have spent a good deal of time writing about the challenges of making a living from the land. During the past two decades my focus has been largely on Africa and other parts of the developing world, but I have kept in touch with farming friends in Yorkshire and followed the news about developments in the countryside, often with a sense of frustration.

Frustration, because it seems that although a great many people are fascinated by food – just look at the viewing figures for programmes like MasterChef – they often know little or nothing about our oldest and most important industry. You could get by without a car or electric lighting, without soft furnishings and fine wines, without newspapers and iPads and trips to Barcelona or Bognor. But however rich or poor you are, whether you live in a city skyscraper or a remote cottage, you must eat to live.

If our distant ancestors were not farmers, they were probably related to farmers and would have had a keen understanding of how their food was produced. That's no longer the case. Many people living in Britain today must go back several generations before they find a member of the family who worked on the land, and their knowledge of country matters comes largely from books, the Internet and television, rather than hands-on experience. No wonder, then, that most people know little about food production and even less about the people who produce it.

While I was researching *Land of Plenty* – I made my way round the countryside in a motorhome, beginning one spring, ending the next – I heard countless stories that illustrate our ignorance. One farmer told me how a customer in a farmers' market in Northamptonshire returned to complain about a chicken he had just bought. 'You've sold me a dud,' he said. The chicken, he explained indignantly, only had two legs. He'd been expecting four, which is how they come in a supermarket pack. It is not just people living in urban areas who know little about farming matters. 'I often give talks to Women's Institutes in villages round here', said a Lake District dairy farmer, 'and I frequently have to explain that you can't get milk from a cow until she's had a calf. Many people don't understand even the simplest things about farming.'

However, this doesn't mean we are ignorant about the countryside. According to the philosopher Roger Scruton, rural documentaries – 'simultaneously users' manuals and dreamers' rhapsody' – had become the most popular form of non-fiction

Beef cattle on summer pasture in Cornwall.

among the reading public by the beginning of the twentieth century. Drop into any good bookshop today and you will still find tables groaning under books about the countryside. A few may be about farming matters, the most notable during recent years being James Rebanks' *The Shepherd's Life*, but the vast majority come under the headings of nature writing, rural travelogue or conservation. Many are suffused with what Scruton describes as a culture of lamentation. They focus on the destruction of our rural heritage, desecration of landscapes and loss of wildlife, with farmers and landowners often cast as villains.

When the tone isn't angry, it tends to be elegiac: the past is another country and superior, culturally if not in an economic sense, to the present. 'The last days of my childhood were also the last days of the village,' wrote Laurie Lee in *Cider with Rosie*, a memoir set in rural Gloucestershire around the time of the First World War. 'I belong to that generation which saw, by chance, the end of a thousand years' life… Myself, my family, my generation were born in a world of silence; a world of hard work and necessary patience, of backs bent to the ground, hands massaging the crops, of waiting on weather and growth; the villages like ships in the empty landscapes and the long walking distance between them.'

You can almost hear the soundtrack: the lowing of cattle, the swish of the scythe, the clip-clop of horses' hooves on rutted lanes. But soon, wrote Lee, the village world would dissolve and scatter, scrubbed clean by electric light and overrun by pensioners. To which could now be added – *Cider with Rosie* was published in

1959 – second-home owners, tourists and shops selling National Trust tea towels and rustic bric-a-brac.

Many of the older farmers I met on my travels could also look back, like Laurie Lee, to a lost way of life. Many told me how their parents had managed to survive and sometimes thrive, without any subsidies, on modest-sized farms with twenty or so dairy cows, fifty sheep, a dozen pigs and assorted hens and geese – something which would be impossible today. These were old-fashioned mixed farms, blending livestock with arable crops, and they provided not only a living but near self-sufficiency in food for the family. I heard stories about ploughing with horses and milking by hand; about trekking cattle and sheep to distant markets; about how they, as children, would catch the blood of pigs whose throats had been slit for the making of black pudding. These stories were always told with relish, but seldom with the sense of regret or nostalgia you find in so much countryside writing. They simply served to illustrate the way farming had changed over the past half century or more.

If you go to an agricultural show or a livestock market the chances are you'll come away with the impression that the farming industry is a bastion of conservatism. You can see it in the way the men dress: the tweed jackets and Barbours; checked shirts with floppy collars, sometimes accompanied by a woollen tie; the rugged footwear and cloth caps. If you listen to the conversations, whether about sheep or politics, milk prices or crop diseases, you may get the impression that this is a tribe apart: dismissive of the metropolitan niceties of political correctness, and firmly rooted, ancestrally and by inclination, to the mores and customs of earlier generations.

However, this veneer of conservatism conceals a remarkable ability to change and adapt, to embrace innovation and think creatively. I am not suggesting this is universally true; you will meet farmers in the pages that follow who still appear to be leading lives little different from those of their forebears. But many are spirited adventurers who are doing whatever needs to be done to survive in a world of uncertainty. Our inconsistent weather, fluctuating commodity prices, livestock diseases like bovine TB, the decision to withdraw from the European Union and its Common Agricultural Policy: these and many other factors have conspired to create an unsettled future for Britain's farmers.

I recently suggested to a group of dairy farmers that if low milk prices are partly a result of overproduction, they should collectively agree to produce less milk. 'But you don't understand,' replied one excitedly. 'It's not in our nature. We are individualistic, temperamental, stroppy. We're not good at cooperating.' Many farmers are some or all of these things, but this is a lazy caricature: John Bull in a rural setting.

Indeed, I can think of no other walk of life that possesses such a diversity of people and enterprises. They range from highly capitalised arable farms in the fertile lowlands, sometimes owned by city investors, to windswept mountainsides where – in the words of the poet R. S. Thomas – an ordinary man of the bald Welsh hills 'pens a few sheep in a gap of cloud'; from ultra-modern indoor dairy units producing millions of litres of milk a year to small, old-fashioned farms making cheese with twenty or thirty cows; from landowners whose families have farmed the same bit of land since Norman times to tenants who have just joined the industry; from estates which use drones and the latest computer-based technologies to small dog-and-stick operations where the most complicated piece of machinery is a battered old Massey Ferguson tractor.

Yet despite their differences and regardless of their economic and social status, they have one thing in common: they provide us with our daily bread and butter, milk and cheese, meat and fish, fruit and salad. At a London demonstration, held in March 2016 and organised by the pressure group Farmers for Action to high-light the impact of low milk prices, many came with home-made placards. One of the more poignant, scrawled on a piece of card-board suspended on baler twine from the neck of a young farmer, simply said: YOU Need A FARMer 3 Time A Day * Breakfast * Lunch * DINNER.

Land of Plenty provides a portrait of a fast changing world at a particular time in our history. The first chapter focuses on a family farm in Leicestershire and introduces many of the key themes – from farm subsidies to animal welfare, the sustainability of our soils to the quality of our food – that recurred throughout my travels. The chapters which follow explore different parts of the farming industry – dairy, sheep, fruit and so forth – in differ-ent parts of the British countryside. I hope the book will provide you with new insights into the people who produce our food and the challenges they face.

1   To the Heart of England

Not long before midnight Mike Belcher knocked on the door of my motorhome and said: 'There's something you might want to see. I'll have to help one of the ewes. She's having triplets.' I followed him to the floodlit lambing sheds and we clambered over the metal sheep hurdles. Mike grabbed hold of a large Masham ewe whose back-end was showing signs of the forthcoming birth and rolled her onto her side. He inserted his right arm almost up to the elbow and felt around inside the ewe. He pulled the first lamb out – glistening and rubbery in the first few seconds of life – and gently swung it up to the mother's head. As she began to lick the lamb, the next one was retrieved from the womb and the process repeated. A few minutes later, the ewe was on her feet and the wobbly-legged lambs began to suckle. The air, fresh and invigorating, smelt of straw and manure.

I told Mike I found the sight of this assisted birth on this clear April night very moving. I expected him to accuse me of being sentimental, but he simply replied: 'Yes, it is, isn't it?' Over the past two months, some 1,900 ewes had lambed at March House Farm, a gently rolling spread of pasture near the village of Great Dalby, a few miles from Melton Mowbray. Now there were just 100 or so sheep still waiting to lamb. Most of the sheep here give birth to twins, but some produce triplets, and when they do the Belchers take pressure off the mothers by removing the weakest lambs and raising them separately. There were about sixty orphan lambs in a large pen on their own. Most were asleep now, but a few were sucking milk from the automatic feeders.

I was amazed that Mike was capable of doing a night's work after such a long day. He had left before 5 o'clock in the morning for Sunday markets in London and returned at 7 o'clock in the evening. 'I don't think of it as working long hours,' he said. 'If I wasn't lambing, what else would I do? Watch television? I'd be bored. I love the nightshifts, especially on a really cold night when the sky is full of stars and there is frost on the sheep's backs.'

There are two reasons why I began my journey here. The first was that I already knew Mike. For over ten years, my family had bought most of our meat supplies – Masham lamb, native breed beef, Gloucester Old Spot bacon, turkeys at Christmas – from his stall at a farmers' market in south-west London. I had even visited him when I was writing about the role that badgers play in transmitting bovine TB to cattle and the fraught and time-consuming business of testing for the disease. I saw Mike as an example of a modern farmer who is making the most of his entrepreneurial skills. Instead of saying goodbye to his animals when they leave the farm gate, like so many farmers do, he adds value by butchering them and selling direct to people like me. Indeed, had he not adopted this strategy, he probably wouldn't have survived as a farmer.

There was another reason for coming to Leicestershire: the county has always been one of the most progressive in terms

Testing for bovine TB at March House Farm.

of its agricultural development. Together with Norfolk, it was the epicentre of the agricultural revolution which took place in the second half of the eighteenth century. Improved methods of cultivation, the adoption of new crops and advances in livestock breeding played a vitally important role in feeding the rapidly expanding urban population. One of the leading figures was Robert Bakewell, born in the Leicestershire village of Dishley. 'By providing meat for the millions, he contributed as much to the wealth of the country as Arkwright or Watt', wrote the agricultural historian Lord Ernle, referring to two of the great inventors during the Industrial Revolution.

Before Bakewell conducted his experiments in stock breeding, most farmers chose animals on the basis of fanciful points of interest, such as the shape of a sheep's horns or the markings on its muzzle. Bakewell, in contrast, focused on the traits which really mattered, such as speed of growth, the weight of the best joints and ability to thrive in difficult conditions. Largely as a result of the scientific approach to breeding pioneered by Bakewell, the weight of calves and lambs sold in London's Smithfield Market rose threefold in less than 100 years.

There has been an ever-increasing trend towards specialisation on British farms. If you travelled around Cambridgeshire, Lincolnshire and East Yorkshire thirty or forty years ago, you would have seen livestock on virtually every farm. Now, you can travel great distances without seeing a cow or a sheep – on one day during my travels in east England I went over 30 miles without seeing any livestock – and mixed farming has become the exception, rather than the rule in some parts of the country. On many arable farms, there is little sense of the past, of the history of land-use and the changing methods of cultivation: huge machines, little labour and copious doses of chemicals are the key ingredients of production today.

In contrast, mixed farming systems, combining crops with livestock, are more complicated affairs and more sustainable in terms of their future, with grass and cereals providing a significant portion of livestock feed and the livestock providing organic manure to fertilise the land. 'We combine the best of the past with the best of the present,' said Mike one day. He probably wasn't thinking of his debt to Robert Bakewell or the great livestock breeders of the past; rather, he was referring to the fact that he still relies heavily on traditional methods of retaining fertility, while at the same time taking advantage of modern technology. Time spent with farmers like Mike is a journey through both the past and the present.

I spent the afternoon before my arrival at March House Farm striding around the Leicestershire countryside. With its gentle undulations, it looked like a rumpled sheet embroidered with brilliant patches of yellow oilseed rape. The hedges were splashed with the white blossom of blackthorn and yellow celandines grew in the soggy runnels beside the roads. The first green buds were just appearing on the oak and hazel but the ash were still in winter apparel, their upturned branches like black fingers against the sombre sky. Although I heard robins and chaffinches, as well as jackdaws and the occasional cock pheasant, there was none of the exuberant birdsong you get on a fine spring day. The ground was still so sodden that few cattle had been turned out to graze from the buildings where they had spent the winter. But at least the countryside was alive with the sound of young lambs – a harbinger, one hoped, of warmer times to come.

When I arrived, Mike's wife Heather suggested I park my motorhome behind the butchery. Once I had plugged myself into the mains I found her in the kitchen, where her granddaughter Florence, one of the cheeriest infants you could ever hope to meet, was playing on the floor with the sort of toys you expect on a proper farm: tractors and trailers and JCBs. We drank strong tea and chatted about the farm.

When Heather was young – she was born here – her father had a small herd of dairy cows, but he eventually gave up milking and concentrated on fattening cattle for the meat market. He had a particular liking for Masham sheep. They are now less popular than they used to be, partly because they are on the feisty side and have a heavy fleece, but Heather and Mike remain devoted to the breed. They also rear pigs in a small way and beef cattle on a larger scale. She couldn't remember how many bulls there were; it was either four or five. Recently, they had to get rid of a Shorthorn. 'One of its testicles was small and the vet said he couldn't back it to perform well,' explained Heather. This is the sort of chatter you get over a cup of tea on an English farm.

I had always assumed that Mike came from a long line of yeoman farmers; he certainly looks the type: tall, broad-chested, ruddy-faced with forearms like hams and a convivial nature. But no, explained Heather, he was brought up in a suburb of West London. His mother used to take him back to the family farm in Ireland during school holidays, and it was his experience there that encouraged him to go to agricultural college, where he met Heather.

As soon as Mike got back from London, he had a quick wander around the lambing sheds, using a metal crook to grab two ewes that he wanted to inspect more closely, then set about emptying the van. While we were taking unsold joints of pork and lamb out of their wrappings and packing them away in the capacious fridges in the butchery, he told me about the farm's recent history. Two years after he married Heather, her father died and they moved into the farmhouse, bought the stock and began paying rent to other family members with a stake in the farm. In those days, they had 180 Masham ewes and a small herd of suckler cows – beef animals whose calves were born in the spring and weaned at the end of summer.

'We managed to scrape a living in the 1980s, but once the 1990s farming depression took hold and prices for agricultural goods fell, we weren't making enough money to service our borrowings, pay the rent and stay afloat,' said Mike. The bank manager was getting tetchy – he was known by local farmers as Mr Sellafield, for obvious reasons – and Mike decided to get a salaried job. For the next ten years or so he was a salesman for a fertiliser company. In the evenings and at weekends he worked on the farm. 'Then in 1999 we heard about a farmers' market in Nottingham and we thought we might as well give it a try,' he recalled. 'We came back after our first day with some cash in our pockets. We suddenly realised that if we could sell direct to the public we might make a proper go of the farm.'

In 2004, the Belchers built a butchery at March House Farm and this now employs two full-time butchers. Two ladies help part-time with the packing. Mike spends much of his time in the butchery and during my week on the farm I often found him making sausages and hamburgers, mostly using recipes devised by Heather. Depending on the time of year, approximately seven or eight cattle, twenty-four pigs and eighty-odd lambs are butchered each month and prepared for sale at six weekly farmers' markets in London and the Midlands. The Belchers also sell livestock direct to Morrisons, which gives a premium for native breed beef, and they sell some of their stock at local auction markets.

Selling at farmers' markets has made a big difference. If Mike sells one of his cattle in an auction market, he gets around £1,000. By doing all the processing on the farm and selling direct to the public, the same animal is worth around £2,000. Take off all the costs – of killing the beast, hanging it in cold storage, butchering it and so forth – and he still makes £500 more by selling direct to the public than he does when selling at an auction. The same goes for fat lambs. These fetch around £70 at auction. Sold in a farmers' market, the legs and shoulders alone are worth well over £100.

Mike and Heather were finally able to buy March House Farm a couple of years ago and they got a bank loan to buy a neighbouring farm of roughly similar size on the edge of Great Dalby. This is where Tom, the younger of their two sons, now lives. The Belchers also have the tenancy of a farm, owned by the Ernest Cook Trust, which covers some 260 acres* at Little Dalby. Their other son, Dan, lives there. At the time of my visit this was where most of the 400 or so beef cattle were to be found, still in their winter accommodation.

Looking back at my notes now, I realise that many of the conversations I had with Mike and his family, and with their young shepherdess and the butchers, touched on issues which would recur time and again during my travels around the country, from animal welfare to the quality of their produce, from the sustainability of the soils to the importance of finding and keeping a committed labour force, from the problems caused by a surfeit of

---

* Like many farmers, I have used both hectares and acres as a unit of measure. 1 hectare equals 2.47 acres.

bureaucracy to the challenges of surviving in a world where the free-market god of cheapness means that the weak, the incompetent and the unlucky are unlikely to survive.

It didn't take long for some of these issues to crop up during my stay with the Belchers. On my first morning, Heather lent me the keys to an old car and I drove over to the farm at Little Dalby. Eventually I found Tom, an immensely tall, cheery, russet-haired figure – these seem to be family traits – mucking out the pigs. 'This looks to me like a nice set-up for the pigs,' I suggested. Tom was forking muck onto a wheelbarrow; the piglets, temporarily released from their pen, were scuttling and grunting excitedly

Gloucester Old Spot piglets on straw.

around the passageway in an old stone-walled shed. There were about three dozen piglets a few weeks old and three sows, each with a recently born litter, in separate pens. A battered radio was tuned to a music station.

Tom stopped forking briefly. 'Well, that's your opinion,' he said. 'But others might say it's a horrible life, being fattened up in concrete pens. It all depends how you look at it.'

'Is that what you think?'

'No, I think it's fine,' he said after some reflection. 'They're well looked after, and after a few weeks the weaners are moved into larger sheds on straw. I'll show you when I've finished this.' As he was spreading fresh straw in the pens he said he thought that large-scale modern pig units were sometimes unfairly maligned. Although some 40 per cent of breeding sows in Britain live outdoors, the vast majority of pigs are fattened for the table in intensive production systems. Many consumers, including most of my friends, I imagine, buy sausages and joints with an 'outdoor reared' or 'outdoor bred' label because they think these systems are kinder. That may not always be the case, suggested Tom. 'There might be times of year, such as winter, when pigs are better off indoors than they are in open-fronted sheds, like the ones we've got here, or outside.'

Most people believe that animals reared outside are better treated than animals reared indoors; and that animals reared under intensive systems of production, for example in indoor piggeries, are worse off than animals reared extensively. When I was writing about British farming some thirty years ago, I was firmly of this belief. However, I now realise this is too simplistic. The mantra of 'outdoors good, indoors bad' may be true some of the time, but not always, depending on a whole range of factors, including species and breed, the nature of the terrain, quality of buildings, the quality of stockmanship and the time of year.

If you go to the website of Viva!, an animal-rights organisation committed to promoting veganism, you will see some distressing videos taken undercover in factory-farming operations. In one of them, a woman asks: 'How would you like it if you were treated like this?' In other words, she is asking you to imagine you were a pig. This sort of anthropomorphism does nothing to advance the cause of animal welfare, although that's not to deny that Viva! has identified examples of poor welfare and even cruelty. A less emotive way of assessing whether an animal is being decently treated has been promoted by the UK Farm Animal Welfare Council, which came up with the idea of Five Freedoms some twenty-five years ago. These are freedom from thirst, hunger and malnutrition; freedom from discomfort; freedom from pain, injury and disease; freedom from fear and distress; and freedom to express normal behaviour. It is worth bearing these in mind as we make our way round the countryside.

As we inspected the rest of the pigs – the Belchers have thirty-odd sows, most being crosses between Gloucester Old Spots and faster-maturing breeds like the Large White – Tom asked if I had been following news about Tesco's labels. Several newspapers had recently revealed that Britain's largest supermarket was using fictitious names evocative of the British countryside, such as Nightingale Farms and Rosedene Farms, on products which included ingredients from countries as far afield as Argentina and Senegal, where welfare and labour standards are inferior to those in this country. The general public, he suggested, were easily duped and their priorities were often curious. For example, customers at farmers' markets were much taken by the idea of Gloucester Old Spot pigs; they liked the sound of this traditional pig and its association with old-fashioned farms. But they seldom asked about welfare, which Tom considered more important than the breed; more important, that is, if you're looking at life from the pig's point of view and not your own.

'What really matters is how the animal is reared, how it's killed, and avoiding stress on the way to be slaughtered,' said Tom. At the Belchers' farm, the pigs are loaded onto a trailer with straw bedding the night before the journey to the abattoir, so they have time to get used to the mode of transport. As it happened, it was three cows – two heifers and a seven-month-old calf – which Mike drove to Grantham abattoir just before dawn the following morning. When we arrived at the abattoir, which was tucked away on an industrial estate at the edge of the town, he ushered the three cattle – or beasts, as he called them – into a pen, where they were inspected by a Spanish vet. She had been at the abattoir since 4 o'clock, checking the welfare of the livestock. I asked why she was here, rather than Spain. For the work, she replied, even though she couldn't bear the weather.

Mike asked if I wanted to watch one of his beasts being slaughtered and we peered through the heavy plastic curtains at the back of the abattoir to observe the preparations. The health inspector and the abattoir workers wore white hair nets, white boiler suits and white gumboots. The men sharpened their knives. A radio blasted out some good rock music. The first beast was encouraged into a metal crate; a man on a gantry lent over the side and stunned it with a captive-bolt pistol. He then pulled a lever and the body slid onto a rack. Chains were wrapped around its back legs and it was winched towards the ceiling. Its throat was swiftly slit. The head was sliced off and the stomach removed. It was then skinned. The journey from life to the chiller took about ten minutes, by which time the two other beasts were making their way along the disassembly line. The veal calf would be delivered back to Mike the following day and the two heifers returned to the farm after a week or so hanging at the abattoir.

I want to know that the animals I eat have been decently kept and decently killed. These were, and I took particular pleasure that weekend in cooking and eating rose veal liver which came from Mike's calf, served with mashed potatoes and a Marsala and onion sauce. I also like the idea that there is hardly any waste. March House Farm uses every bit of the animal, with the exception of the lungs and windpipes. Bones are used to make fertiliser, pigs' eyes go to local schools for biology lessons, hearts are used in faggots.

When we got back to the farm, Mike introduced me to Pete Osborne and Paul McDonald, who were in the butchery preparing for the next day's farmers' market. Both had long experience of working in supermarkets and high-street butchers before coming to March House Farm. They clearly enjoyed themselves here, and whenever I dropped by the butchery – say, to get bacon for my breakfast – there was much light-hearted and sometimes fruity banter. They liked the fact that this was a traditional butchery, requiring a range of skills. Today, they were preparing four quarters of beef, three lambs and six sides of pork.

'The meat you get here is a better flavour than the meat you get in most supermarkets,' suggested shaven-haired Pete, the more garrulous of the two. This was partly because of the breeds used on the farm, partly because of the way the animals have been fed, killed and hung. It was particularly important that they weren't stressed at the time of slaughter. 'If they are, you can immediately tell from the meat,' said Paul. 'Meat from a stressed pig will be floppy, from a beef animal a bit darker. When you try to cut into it, it's like carving through wood.' They took a dim view of the mode of slaughter for the Halal and Kosher markets. Instead of being stunned first, the animals have their throats slit. 'I have seen a video of it, and it's horrible,' said Pete.

In the great medieval poem *Piers Plowman*, the land of the living is referred to as a 'fair field full of folk'. In those days, most of the population would have worked in the fields and made a living from the land. During the Industrial Revolution, there was a great exodus from the land as people moved to the towns and cities, and fled rural poverty, in search of a living. By 1851, half the population lived in seventy towns which had over 20,000 inhabitants each, but there was still a significant farm workforce of 1.7 million men and women. By the end of the Second World War, the number of farmworkers in the UK had fallen below 1 million and by 2012 there were just 172,000, which included 67,000 'seasonal, casual and gang labour', as they are described in official statistics. Most of the latter are migrants from Eastern Europe.

The average age of farmers in the UK is said to be fifty-nine; this is exactly the average age of farmers worldwide, according to the United Nations. It is one of those statistics that sounds compelling and distressing – if farmers are so old, how are they going to feed us in future? – but it is also misleading. It may be true that the average age of the farmers whose names are on title deeds or tenancy agreements is fifty-nine; but that doesn't mean that the industry is run by a rustic gerontocracy. Over the last few years, I have been to a great many agricultural shows and I have been heartened by the number of young people – not just farmers' children but others interested in farming – in attendance. I imagine the Belchers are typical of many medium-sized farms, with Mike and Heather in their sixties, two sons in their early thirties and one full-time farmworker in her early twenties.

A tall, willowy woman with long blonde hair, the palest of pale blue eyes framed with mascara and brightly varnished fingernails, Jane Riley was perennially cheerful and always happy to chat when I visited the lambing sheds. Yesterday, she explained with a broad smile when I first met her, was the first day in over a month when she had found time to leave the sheds. She had gone out to the fields with Dan, Mike's elder son, to look at a flock that had lambed some weeks ago. Ten of the sheep were hers.

Mike Belcher lambing at night.

'It's just a hobby really, keeping a few of my own sheep,' she said. She produced a phone from her jeans and flipped through photos of her sheep. They were Scottish half-breeds which had been put to either a Suffolk or a Charollais tup. It takes a while to get used to the complicated breeding systems of the sheep world, but I was beginning to get the hang of it. The principle is often the same: a hardy hill sheep – such as a Swaledale or a Cheviot or a Masham – is crossed with a lowland breed, such as Bluefaced Leicester. Their progeny are then crossed for hybrid vigour – crossbred individuals tend to have superior qualities to both their parents – with another lowland breed to produce the lambs we eat.

So why had she chosen these particular breeds? 'Oh, I just love the way they look,' she said.

Jane resumed her daily tasks, moving ewes which had lambed the day before from single pens back to a larger group. She picked the lambs up and encouraged the ewes to follow, bleating *maa-maa-maa*. 'I get bored easily, and I'm more of a practical person than a paperwork person,' she said. 'That's why I love farming. I love it here because they've got everything – sheep, pigs, cattle, cereals. Every morning, I wake up and I say to myself: "Yes! I've got to go to work!"' She felt that she was an equal here, which says much about the ethos of the Belcher family.

Early one morning I found her performing some intricate tasks on newborn lambs. 'I'm castrating them,' she explained. This involved using a piece of equipment like a pair of pliers to place a rubber ring at the base of a lamb's pouch, which is what she called the scrotum. She then repeated the exercise about a third of the way down its tail, gave it a jab against infection, and dabbed some surgical spirit around the navel. All this took a matter of seconds.

'Have a feel,' she said as she was doing another lamb.

'A feel of what?'

'His balls.'

She invited me to castrate the next lamb. Once you've placed the ring on the pouch, you need to make sure that you haven't trapped the lamb's balls or its nipples. Over time, the balls will atrophy, as will the lower two-thirds of the tail. It is important to get that right as well, she explained, because you could risk being fined if you didn't. Two generations ago, farmers often used to

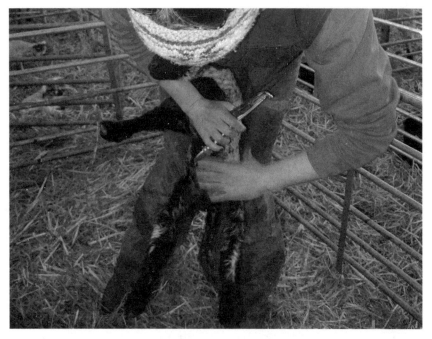

Jane Riley vaccinates a lamb against infection.

castrate their lambs by cutting open the purse with a razor blade, sucking the balls out and biting off the tubes. Hence the old adage: you're not a proper shepherd unless you've done it with your teeth.

A couple of years ago I suddenly found myself with time on my hands when an assignment in Africa fell through. I decided to head down to the West Country, where I dropped in to see some old farming acquaintances and attended a sheep fair. I also took the opportunity to stretch my legs and enjoy the first warm flush of summer. I recall, in particular, heading out on foot in search of a pub meal in the voluptuous dairying countryside near Tavistock, in South Devon. The hedges were so high, and in the valley bottoms bowered by a canopy of trees, that I felt as though I was walking down a long green tunnel, with occasional gateways providing a passing window on the fields beyond. Oak, hazel, field maple, ash, blackthorn and elder were intertwined like hand-holding lovers; honeysuckle, foxgloves, ivy and goose grass clambered up their mossy limbs; late-flowering bluebells were a washed-out blue in the shade of the unfurling fronds of fresh bracken. The air smelt of cow dung, honey and fresh salad. A linnet sang lustily from an elder bush near the River Tamer and swallows skimmed over the green, cow-patted pastures.

In *Afoot in England*, W. H. Hudson described his wanderings about the English countryside in the early years of the twentieth century. Like many nature writers, he was filled with nostalgia for earlier times and he had an intense dislike of many aspects of modern life; of 'parasitic and holiday places', for example, and the 'city-sickened wretches' who inhabited them. He was also disenchanted with the landscape of South Devon: 'There is too much green, to my thinking, with too much uniformity in its soft, bright tone.'

Unlike Hudson, I love the South Devon countryside and the way human history is written in its contours and vegetation. The hedges here provide a diorama of past land-use and human endeavour. Some of Devon's 30,000 or so miles of hedges are over 800 years old, and most of them were created by farmers. They were planted to restrain livestock and they still perform the same function today. Some hedges were made from remnants of ancient woodlands; others were deliberately planted, often on earthen banks.

Elsewhere in England, most of the hedges we see today owe their origin to the great enclosures. Between 1600 and the First World War, Parliament passed over 5,000 Enclosure Acts. These privatised areas that had formerly been under common ownership, depriving the peasantry of the strips of land where they grew their crops. Prior to the enclosures, livestock were able to wander freely after the cereal harvest, thus making the growing of winter crops difficult if not impossible. The enclosures meant that farmers were able to grow turnips, clover and other feed crops after the harvest and abandon the habit of slaughtering most livestock at the onset of winter. The enclosures led to higher crop productivity, the development of improved livestock and the more efficient use of labour. Without enclosures, and without hedges, agricultural improvement would have been impossible.

Inevitably, England's peasant farmers lost out as common lands were appropriated for private use and many headed for the cities or became farm labourers. Those who remained in the countryside frequently endured terrible hardship. According to the historian G. M. Trevelyan, farm labourers in the south were often forced to take their wages 'in bad corn and worse beer'. In the 1830s, starving labourers burned corn ricks and marched about in a riotous manner; three were hanged and over 400 were transported to Australia as convicts. You might think, when reading Thomas Hardy's *Far from the Madding Crowd*, that farm labourers were a bucolic and boozy bunch wanting for little other than a decent education and a supply of shampoo. It wasn't like that. Indeed, Hardy himself knew a boy in Bockhampton, the Dorset village where he lived, who starved to death while looking after sheep. He had been trying to survive on raw turnips.

Yet peasant life before the enclosures must, at times, have been a grim affair. There were certainly some good periods, when the peasantry thrived, but it wasn't always like that. For many it was a life of bare sufficiency and frequently want, with little or no surplus to sell and no means of buying goods and services. Virtually every peasant farmer I have met in Africa – and I don't imagine the peasant way of life in medieval England was much different in terms of the things that really matter: shelter, food, health – has aspired to go beyond subsistence farming, to join the market economy, to benefit from the fruits of modernisation. It always gives me great pleasure when I visit a remote African village and see farmers using mobile phones to find out the price of maize or millet in distant markets, before deciding whether or not to harvest their crops.

If reading W. H. Hudson leaves me with a sour taste in the mouth – it's not just his snobbery which grates, but his romantic portrayal of peasant life – the opposite is true of William Cobbett, whose *Rural Rides*, published in 1822, is one of the great works about the English countryside. Cobbett railed against the cruelty of the landowning classes and the wretchedness of farm labourers, but he was not so blinded by rage that he failed to recognise good things when he saw them. Intriguingly, he found that the richer the land and the more lacking in woods – 'that is to say the more purely a corn country' – the more miserable the farm labourers. In contrast, he noted the happy state of labourers in Worcestershire, where livestock and crops were reared and grown together.

I was reminded of Cobbett's praise of mixed farming when I spent a day with Dan Belcher, who invited me to join him when he was taking a consignment of ewes and lambs to Blaston, in south-east Leicestershire, where the Belchers have a grazing agreement. This all came about because his father fell into conversation one day at Chiswick Farmers' Market with Hylton Murray-Philipson, the owner of the estate. A successful entrepreneur, investment banker and conservationist, Hylton told Mike that he was worried about the state of the soils at Blaston. At the time, most of the estate was devoted to arable crops and devoid of livestock. The soil was exhausted, the bill for agrochemicals seemed to grow ever larger, and the fields were suffering from an infestation of black grass, a vigorous weed which is a symptom of poor land management and one of the plagues of modern agriculture. Hylton told Mike that he was thinking of introducing livestock and adopting various other measures to improve soil fertility. He wondered whether the Belchers could help.

When Dan and I arrived, a pallid sun was struggling to break through the bruised clouds, but the countryside still looked lovely, with a winding stream and a scattering of handsome buildings in the valley bottom, and irregular fields rising up to gently undulating hills blurred with tall copses of ash and other native trees. 'If you look at what's happening over much of England, farmers have been raping the goodness out of the soil,' suggested Dan as he decanted the ewes from the trailer. Such strong language, coming from such a charming and articulate character, took me by surprise, but after spending some time at Blaston – I returned a couple of days later to talk to the land agent and farm manager – I understood exactly what he meant.

For many years, Blaston, like so many other farms in lowland Britain, had adopted a short, repetitive crop rotation, with two high-value crops, wheat and oilseed rape, following one another year after year. This simple way of farming was driven by economic considerations – it made sense when prices for cereals and oilseeds were high – but it was environmentally reckless, although most farmers didn't realise this at the time. Now, with the help of the Belchers, Blaston was undergoing a process of transformation, which involved reducing the area devoted to cereals, increasing the area under grass and getting fertility from livestock rather than out of a plastic bag.

Fertility is the key to all life. It is no coincidence that early civilisations were founded on the banks of rivers which flooded every year. Some 500 years before our ancestors built Stonehenge, the Ancient Egyptians established the first nation state. Every year, floodwaters from Ethiopia and Sudan deposited a thick layer of black silt over the parched earth beside the Nile. As the waters receded, farmers planted their seeds in the fertile soil. They grew wheat for bread, barley for beer, cotton for clothes, vegetables, date palms and olives. They might not have understood the precise chemical composition of the silty floodwaters, but they knew that the inundations made their soil fertile and they attributed their good fortune to the god Hapi, a fat little figure – like the sort of character you see propping up bars in country pubs – who represented abundance. As the Greek historian Herodotus wrote, Egypt was the gift of the Nile.

Naturally fertile soil provides all the elements needed for healthy plant growth, including essential nutrients, trace elements and water. A healthy soil is rich in organic matter, which helps to retain moisture and creates the perfect medium for roots to spread and develop, as well as for living microorganisms. Earthworms, bacteria, fungi and all manner of other organisms help to support plant growth and recycle nutrients.

One of the great advances in British agriculture was the Norfolk four-course rotation, which was developed in the late seventeenth century. Under this system, farmers grew wheat in the first year of the rotation, turnips in the second and barley, undersown with clover and ryegrass, in the third. During the fourth year, the clover – a plant which has the ability to take nitrogen from the air and 'fix' it in the soil – and ryegrass were either grazed or cut as animal feed. The turnips were also used to feed cattle and sheep in winter. The livestock, in turn, provided the manure which restored the fertility of the soil.

The introduction of livestock, some owned by the Belchers and a new herd of South Devon cattle belonging to the farm, and a range of innovative agronomic practices have led to significant improvements in the quality of the soil at Blaston. When I first met Hylton Murray-Philipson, over lunch at the Royal Academy

These native breed cattle will soon be ready for slaughter.

in London, he talked about the difference between fertile soils and soils exhausted by decades of cultivation and heavy use of chemicals. If you went into one of the woodlands on his estate you could drive a spade into the soft compost-rich earth like slicing into butter; try it on one of the old wheat and rape fields and it was like chipping away at concrete. His aim now was to transform the latter into something approaching the former. Exactly a year after I first visited Blaston, I returned to spend a day with Hylton, his land agents and a team of soil scientists. More about that in Chapter 8, which focuses on restoring the health and biodiversity of our arable systems.

After we left Blaston, Dan and I made our way home via the delightfully named estate of Carlton Curlieu, the seat of

Sir Geoffrey Palmer. The Belchers rent 200 or so acres of pasture here on which they graze up to 700 sheep. We drove slowly over the undulating fields, remnants of medieval ridge-and-furrow ploughing, and on a couple of occasions Dan jumped out to chase a lamb which was on the wrong side of a fence. In the parkland in front of the magnificently gabled and much chimneyed seventeenth-century manor house we came across a dead lamb. 'It happens,' said Dan. 'It probably got lost from its mother.'

I asked whether he had any views about the EU referendum, which was less than three months away, and he told me a story about a meeting organised by the local branch of a large high-street bank. He had been invited along with tenants and owners from both small and large enterprises. The big estates were particularly worried about the idea of 'Brexit', presumably because they feared the loss of subsidies, which would almost certainly translate into falling land prices.

But what about you, I asked Dan.

'In New Zealand,' he replied, 'when subsidies were abandoned overnight, only a very small percentage of farmers went out of business. They had to become better farmers, rather than carry on doing the same as the previous generations. That's what I said when it was my turn to speak at the meeting. We should do the same here.'

Farmers here needed to become more entrepreneurial, said Dan. That could involve a whole range of activities, from selling farm produce direct to the public to setting up bed-and-breakfast businesses, renting farm buildings as offices, adding value by transforming milk into cheese; indeed, doing anything which increases the earning potential of a farm.

'Our business is not based around getting the Single Farm Payment,' continued Dan, 'and I'd be happy to do away with it.' The Single Farm Payment, now known as the Basic Payment Scheme, provides all active farmers with an annual payment based on the area they are farming. The larger the holding, the more they get. For the Belchers, this amounted to around £40,000 in 2016. Some of the largest landowners in the country get more than £1 million a year. At the time, none of these payments were under threat. Opinion polls suggested that the majority of voters would opt to stay in the EU; the prevailing subsidy system would be maintained, not least because the politically powerful French farm lobby, in particular, would refuse to countenance reform.

Many of the farmers I met who were keen to get out of the EU, and voted for Brexit, complained about the amount of time, and sometimes financial resources, they had to spend complying with regulations. On my last evening, Mike and I went into Melton Mowbray to buy fish and chips before joining Heather at Little Dalby, where she was babysitting for Dan. 'You should ask her about all the bureaucracy she has to deal with nowadays,' suggested Mike.

We ate at the kitchen table and washed down our battered cod with bottles of cold beer. 'There's so much ticking boxes simply for the sake of it, and filling in forms for the sake of it,' said Heather when she had finished her supper. Back in the 1980s, she worked as a farm bookkeeper for some twenty farms. Now, she spends at least four hours every day just keeping the accounts straight for the farms owned and managed by her family. 'If you get anything wrong, they can take some of your Single Farm Payment,' she explained. The subsidy comes with strings attached, known as cross compliance. In return for payment, the farmer must agree to actively farm the land, keep records related to every aspect of the enterprise, observe animal and plant health guidelines, fulfil various environmental tasks and so forth. If you think about it, this is perfectly reasonable: why should farmers get something for nothing?

Heather told me about a complaint she had recently received from the local council's rights-of-way officer. Under cross compliance they have to keep bridleways and footpaths clear of brambles and undergrowth. They also have to make sure that the ground is in good order. To do this, they recently laid down some hardcore – old crushed concrete – on a muddy and rutted bridleway. Shortly afterwards, they received a letter from the rights-of-way officer saying they had 'adulterated the public highway' and that this was dangerous. If they failed to rectify the matter, they would be taken to court and reported to the Rural Payments Agency (RPA), which could then fine them 3 per cent of their Single Farm Payment. 'That's typical of what happens,' said Heather, adding indignantly: 'And we were trying to make good a poor pathway!'

I mentioned that I had stayed for a night on a farm in Northamptonshire immediately prior to my visit here. The farmer told me that upwards of twenty-six different organisations have a right to come onto his farm, uninvited, to inspect one thing or another – 'and they all earn a living out of it'. If you make a mistake, he said, it is always your fault; if they make a mistake – the RPA or whoever – it is never theirs.

Yes, all quite true, said Mike. Just last week, an inspector from the RPA had come, unannounced, to check on their sheep and make sure all their on-farm sheep movement records were up-to-date. There was nothing unreasonable about this, but it did mean that Tom had to abandon his morning tasks and shepherd the civil servant around the farm. Although Mike was strongly in favour of leaving the EU, he recognised that this wouldn't herald an end to farm bureaucracy: one complaint you will hear from our own farmers is that government departments in this country are more eager to enforce regulatory systems than those of other EU countries. But, like Dan, he believed that leaving the EU would encourage farmers to become more entrepreneurial and more inventive. This, he reckoned, could only be a good thing.

Shortly after I left Leicestershire, I headed south-west to the Cotswolds. I stayed in a small campsite on the edge of the village of Great Rissington, which possesses everything you could want of a small English village: a pub with good food, a green grazed by sheep, a medieval church and a bus shelter with an advertisement for a professional mole catcher. From time to time I would chat with the owner of the campsite, and I heard a story that was repeated almost everywhere I travelled. When he was young, he said, there were four farms in the village. Now there was just one. One farmer's wife told me that there used to be fifteen farms in and around her Gloucestershire village when she was a girl, some sixty years ago. Now there are just three. Not only are there far fewer farms than there used to be in the Cotswolds, but far fewer farmworkers, a pattern repeated across lowland England.

These trends – larger farms, fewer farmers and fewer farmworkers – will almost certainly continue, so the obvious question to ask is: who will the survivors be? The subsidy system has helped to keep in business many farmers who would have gone to the wall otherwise. Now that we are leaving the EU, it is highly unlikely that the support system will survive in its present form beyond 2020. I imagine there will be three main groups of survivors. One will be the most efficient operations, which will often be the largest and most heavily capitalised. Another group will consist of farmers who will receive financial support in return for providing public goods, for example by protecting watersheds and wildlife, or looking after some of our great landscapes. The third group will consist of people who are adding value to the raw materials they produce: people like the Belchers, who you have just met, or the Blands in Cumbria, whose story of rural redemption opens the next chapter.

2 All Churned Up

After the 2001 foot-and-mouth crisis, when over a million livestock were slaughtered in Cumbria as a measure to eradicate the disease, the Lonsdale and Lowther Estates, which own a large chuck of the Lake District, organised a meeting for their tenant farmers. It was suggested that instead of restocking with high-yielding cows and doing what they had always done – churn out large quantities of ordinary milk – they might consider doing something different. One family, the Blands, took this advice; the rest ignored it.

I first met Stephen Bland when he was rolling spring barley on a blustery morning towards the end of April. I walked across the field and when I reached him he turned off the tractor engine. I told him why I wanted to see him and asked if he had time to chat. 'Come back tomorrow,' he said cheerily. 'You know, fifteen years ago everyone laughed at us when they saw what we were doing – making ice cream. They said we would go bankrupt.'

A balding, fit-looking man in his fifties, Stephen was dressed as though he were auditioning for the part of a scarecrow. Indeed, he used his clothes, tugging at an ancient sweater so full of holes that it might have been pulled out of an archaeological dig, to stress the importance of keeping costs down on the farm. He laughed a lot when he spoke and his irrepressible nature was in marked contrast to the calm and carefully considered manner of his wife Claire, who joined us the following day when we met in the farm tearooms.

Stephen was a man with a story to tell, and the story began in East Lancashire. He didn't come from farming stock – his father was a builder and his mother worked in the cotton mills – but from an early age he helped out on a local farm belonging to the Bridge family, who he talked about with great affection. They had some thirty dairy cows, as well as pigs and sheep and turkeys for Christmas. They were traditional small-scale mixed farmers and they bottled the milk themselves and sold it in the surrounding villages. 'In those days, farms always had three or four of us lads knocking around,' recalled Stephen, 'and on Saturdays Mrs Bridge would give us home-made toffee, made from the treacle they fed the cows.' This sort of thing could never happen nowadays. For one thing, there are hardly any old-fashioned producer-retailers left; and for another, health and safety regulations would never allow it.

Stephen's first serious farm job was in Barrow-in-Furness, milking 250 cows three times a day. The owner was a good farmer, but eccentric, and liked to round up the cows in an old Mercedes. A year or two after he left there, Stephen met Claire at an event organised by the Young Farmers' Club – 'the best dating agency there is' – and they rented a smallholding. Eventually, in 1999, they were offered the tenancy of Abbott Lodge Farm, which occupies around 230 acres of flattish land a few miles south of Penrith.

'When we came to see the farm, the agent told us not to be put off by the terrible state it was in,' explained Claire. The Blands had to remove over 1,000 tonnes of muck from the yard before they could use the farm buildings. The roof over the dairy parlour had partly collapsed. There were twenty-seven broken pipes and most of the drains were blocked with plastic bags. There was no electricity in the outbuildings. Claire flipped through the photographs in a small album as she described the state of the farm. Two stick in my memory: one was of the former tenant, a portly-looking character dressed in tweeds like a country gent, smiling for the camera; the other was of a dead calf lying on several feet of muck.

In September 1999, Stephen and Claire walked their herd of 100 cows down the road to Abbott Lodge Farm. 'What followed was the most uncomfortable time of my life,' said Stephen. Milk prices fell and one of his best friends was killed in a farm accident. On 28 February 2001, foot-and-mouth disease was detected on a farm at Longtown, near the Scottish border. Some three months later, a foot-and-mouth lesion was found on a cow at the Blands' farm and they were told that their entire herd had to be slaughtered.

'I've still got a whole box of papers about what happened then, but I can't bear to open it, even now,' said Claire. There were no government-supplied marksmen available to kill the cows, so Stephen rang the then head stalker for the Lonsdale and Lowther Estates. 'At first, he said he just couldn't do it. But ten minutes later he rang back and said he'd come.' The Blands' two young children were showered, dressed in clean clothes and taken out of the front door of the farm. Then, under instructions from a vet, the cows were shot in the field, one by one.

'After something like that happens, you could either sit down and say: poor me,' reflected Claire. 'Or you could get off your arse and do something about it. That's what we did.' Using compensation money for their slaughtered herd, they bought sixty Jersey cows from a farm in Cornwall. They also took out a bridging loan from the bank to convert an old shed into tea rooms. 'Everybody thought we were mad,' said Stephen. 'Can you imagine – making ice cream in Cumbria!'

Today the farm is a popular tourist attraction and an important employer of local labour. When I was there it was a cold and miserable day, but there was still a steady trickle of customers, including a couple of adults with learning difficulties who'd come with their carers for tea and cake. During recent years, the Blands have built their herd up to 300 pedigree Jersey cows. Some of the milk is used to make ice cream and the rest goes to a dairy in Scotland. 'There's plenty of farmers round here who are only getting 13 or 14p a litre for milk from their black-and-white cows,' said Stephen. 'Well, I'm getting 37p a litre.' He could have added: and nobody's laughing at us now.

The reason why Stephen is getting such a high price is because he is tapping into a speciality market for Jersey milk, selling to Graham's Dairy at Bridge of Allan. Buy a litre of Graham's 5 per cent fat 'Gold Smooth' milk in Waitrose and it will cost you £1.10;

a litre of standard 3.6 per cent fat whole milk – this is what most of Stephen's neighbours are producing – will set you back just 44p.

I was brought up on milk. My father and I drank a glass with every meal. We poured it on cereals, on puddings, on fresh fruit. When I was working on a farm, I used to take two pints of milk laced with Camp coffee to drink with my lunchtime sandwiches. And when I was a student I marched in protest against the government's decision to withdraw free milk for schoolchildren: 'Mrs Thatcher, milk snatcher!' we chanted tunelessly before lighting another fag and heading to the pub.

No other animal product is as versatile as milk, and I am as fond of its manufactured products – butter, cheese, crème fraîche, yoghurt, ice cream, junket – as I am of the liquid. It has other uses too: Cleopatra is said to have taken a bath each day in the milk of 700 asses; Queen Elizabeth I apparently washed her face in cow's milk. And in the Old Testament, God led his chosen people, the Israelites, not to a land flowing with beef and pickle, or bread and cheese, but milk and honey, a symbol of fertility and abundance. Milk's many uses and virtues make the current state of the dairy industry seem all the more perplexing.

When I was growing up, you would see all sorts of native dairy breeds grazing in the meadows: Ayrshires, Guernseys, Jerseys, British Friesians, Dairy Shorthorns. Nowadays, the vast majority of dairy cows are high-yielding Holsteins or Holstein crosses.

In the 1960s, there were an estimated 150,000 dairy farms in the UK. There are now just over 13,000. In 1965, the average herd size was twenty-six. It reached seventy-five by the year 2000 and it is now 130. Change has been particularly dramatic in the last two decades, with the number of dairy cows falling by 27 per cent and the number of producers by 61 per cent. During the same period, the average milk yield per cow rose by 46 per cent: a triumph of modernisation, when judged purely in terms of efficiency and productivity.

Many farmers, like the family Stephen Bland worked for as a teenager, used to deliver their own milk, or sold it to companies who delivered it to the public. Indeed, as late as 1995, doorstep delivery accounted for 45 per cent of household purchases in England and Wales, compared to 3 per cent or less today. Between 1933 and 1994, the Milk Marketing Board acted as buyer of last resort for British milk, thus guaranteeing a minimum price, which made it easy for farmers to plan for the future: they knew what they would be getting, and fluctuations were relatively modest. The process of deregulation, which began under a Conservative government in the late 1980s, led to the eventual demise of the Milk Marketing Board and greater competition. Good for consumers, perhaps, but not necessarily for producers. A chart displaying the fluctuations in farm gate milk prices now looks like a child's outline of the Cairngorms, rather than the Fens.

Two years ago, the Blands' farming neighbours were getting 37p a litre for the milk from their black-and-white cows, as Stephen called Holsteins and Holstein–Friesian crosses. They were prospering; they had money to spend on new equipment and foreign holidays and new kitchens. A couple of lactations later – the cow's cycle is not much different to a human's – low milk prices meant that many feared bankruptcy and some had already gone out of business. Between June 2013 and June 2016, over 1,000 dairy farms closed in England and Wales, with Cumbria among the worst affected counties. Little wonder, then, that dairy farmers have been at the forefront of protests against falling farm prices.

In March 2016, I attended a demonstration in London organised by Farmers for Action. About 1,000 people gathered in a square just off Pall Mall and before the speeches I wandered among the crowd and listened to the farmers' grievances. Many of them railed against what they saw as the exploitative behaviour of supermarkets and dairies. Supermarkets dominate milk sales and they often use milk as a loss leader; this leads to intense price competition and, inevitably, poor returns for producers. Not that dairy farmers are all treated the same way. Some of the farmers I met on the march had 'aligned' contracts with supermarkets and they were getting around 27p a litre for their milk, several pence more than the cost of production. However, the majority here were being paid less than the cost of production, with some losing as much as 10p on every litre.

Nearly all the farmers I met agreed that we are producing too much milk in this country. Export markets had been hit by declining demand from China and by EU sanctions against Russia, following the latter's annexation of Crimea and interference in

Ukraine. In the old days, the Common Agricultural Policy (CAP) might have solved the problem through intervention buying: this was why we ended up with butter mountains and milk lakes back in the 1980s. But this sort of support is no longer available. Instead, farmers need to decrease production. Some of the smaller operators I spoke to said this should be the responsibility of larger farmers; the latter suggested that the least efficient should be the ones to leave the industry.

Two of the most efficient dairy farmers I met had large indoor dairy herds in the Derbyshire Dales. I was taken to see them by an old friend of Mike Belcher when I was staying in Leicestershire. Brian Hesford picked me up on a gorgeous spring day, the sky a cloudless azure, the hedgerow birds singing fit to bust, the farm-yard alive with the sound of bleating lambs. It took us about an hour and a half to reach our destination and Brian seemed to know every farm we passed, having spent most of his life supplying ser-vices to dairy farmers across a broad swathe of the Midlands. That one is a fine farmer, he would say; and that one – you see, the farm with dilapidated barns – he's been banned from keeping livestock for welfare reasons; this one on the left used to be a dairy farm, but they've just gone out of business; that one, lovely family, I went to the son's wedding. 'It's really hard to see hard-working people being kicked about by supermarkets and dairies,' said Brian as we began climbing up into the good limestone country, all dry stone walls and leafless ash trees, beyond Ashbourne. 'These people are my friends, not just people I work with.'

We chatted for an hour or so to Tom and Sue Flowers in the kitchen at Old House Farm before Sue showed me round the cows. When Tom was born here in 1947, this was a mixed sheep and dairy farm with thirty-five Ayrshire cows. There are no sheep now, just 270 Holsteins kept indoors all year round. The Flowers had adopted every efficiency possible, yet they were struggling to keep their heads above water. 'It's a real challenge just paying the monthly bills,' said Tom morosely. When I asked what needed to be done, Sue said farmers should be paid the cost of production for their milk plus a certain amount extra.

Sue Flowers with her indoor dairy herd.

After a pub lunch we visited Adam Sills, whose 260 Holsteins were producing over 3 million litres of milk a year at Ash Tree Farm, near Rodsley. As we drank tea, made by his teenage daughter, Adam explained that falling milk prices meant that his business had made £300,000 less in the current year than the last. He wasn't blaming Arla, a farmers' cooperative, who had reduced the amount they paid him from 30p to 20p a litre, or Asda, who sell his milk, but global milk prices. 'We farmers are doing too good a job, that's the problem. There is way too much milk, so we're all guilty.'

I asked whether he, like Sue Flowers, thought that farmers should be paid the cost of production plus a bit extra. He shook his head. 'Just imagine if JCB was churning out seventy-five diggers a day and the building industry suddenly went down the Swanee and didn't want them. It would be no good them calling for support from government or anybody else. Well, the same applies to us. There needs to be either a cull of farmers or a cull of cows.'

Roughly speaking, the dairy industry falls into two distinct camps: about a quarter of farmers adopt low-cost production methods; the rest – like these Derbyshire farmers – aim for much higher yields and depend on more intensive methods of production, involving considerable expenditure on bought-in feed. When milk prices are low, as they were in 2016, the highly capitalised intensive systems fare badly; in contrast, farmers adopting low-input systems, largely based on the efficient use of pasture, can still make a reasonable living.

One of the reasons why we are producing too much milk is

because we have created a breed of cow, the Holstein, whose sole purpose is to produce as much milk as possible in as short a time as possible, hence the huge udders. 'Try to get a Holstein to survive on grass, and it would soon die, it'd be like putting diesel in a Formula 1 car,' one farmer told me. 'They need cosseting and a complicated high protein diet.' Holsteins and Holstein crosses now produce 90 per cent of our milk.

Traditional native breeds, in contrast, fare well on grass during the summer and a mix of hay and silage during the winter, with a little protein-rich feed. The Blands' Jersey cows – doe-eyed and not much more than waist-high to a cowman – are hardy, relatively low maintenance and produce around 6,300 litres of creamy milk a year, compared to over 11,000 litres for intensively managed Holsteins. Drink a glass of Jersey milk and it feels like a meal; Holstein milk is closer to white water. Jersey milk also makes excellent ice cream, some thirty flavours of which are on sale at the Blands' farm shop and tearooms.

As we headed back to Leicestershire, I told Brian Hesford that I was amazed that so many dairy farmers wanted to continue when they were losing so much money. 'Look at it from the farmer's point of view,' he said. 'He's always been a dairy farmer, and so have the generations before him. If he stopped now, he thinks his great-grandfather's ghost would rise from the armchair and bash his brains out with a poker.' In any case, most farmers love farming; they wouldn't know what else to do.

What is this life if, full of care,/ We have no time to stand and stare./ No time to stand beneath the boughs/ And stare as long as sheep or cows, wrote W. H. Davies, the Welsh tramp poet, a few years before the First World War. Of course, the cows which Davies was thinking about were outdoor cows, quietly ruminating in green pastures: a poet's summer cows, not cows housed indoors in winter or kept indoors and 'zero-grazed' all year round.

The vast majority of cows in the UK are grazed extensively, spending six months or more of the year outdoors, during which period they only come inside to be milked. During the wetter, winter months, when the grass stops growing, they are mostly housed indoors. A relatively small number of cows spend most or all of the year outdoors. This is known as the New Zealand system and is only suitable for the hardier breeds of dairy cow in the warmer, drier parts of the country. During recent years there has been an increase in the number of herds that are kept indoors for most or all of the year, although some may spend time as dry cows, prior to calving, grazing outside. The practice of keeping cows indoors for such long periods is widely frowned upon by animal welfare organisations, and by some supermarkets too. Waitrose, for example, guarantees that all the milk it sells comes from cows which have been grazing outdoors at least 120 days each year.

But what exactly constitutes good welfare from a cow's point of view? To help me answer the question, I went to see Roger Blowey, a Gloucestershire vet and expert on the welfare of dairy cows. When I arrived at his home in Minsterworth, a village to the west of Gloucester, he was proofreading the latest edition of his book, *The Veterinary Handbook for Dairy Farmers*, first published in 1985. While he was making coffee, he said: 'We need to dispel some of the myths about welfare in dairy farming. It's often claimed that animal welfare is better on smaller enterprises, but all the evidence suggests that isn't the case. Larger dairy farms often have much better welfare than small farms.' Roger had given evidence in court as a veterinary expert, both for the prosecution and defence. Poor welfare, in his experience, was most frequently associated with small, undercapitalised farmers.

One of six children, Roger was brought up on an 80-acre mixed farm near Tavistock. After qualifying as a vet, he has spent most of his working life in Gloucestershire. 'My main interest is how you can create an environment which is economically productive, and in which animals thrive,' he said. I asked whether he had noticed any significant changes in the welfare of farm animals. 'I think most farm animals are mind-bogglingly better off, in terms of their welfare, than they were when I began my career,' he replied.

There were two main reasons for this. For one thing, farmers are now much better trained in animal welfare and the treatment of disease than they were in the past. For another, larger units – and the trend has been towards larger units – can afford more sophisticated equipment, which allows them to monitor virtually every breath an animal takes. 'For example, on the farm I'll take you to today, they recently spent over £20,000 on a cattle crush which enables staff to examine the cows' feet and treat lameness. A small farm simply couldn't afford that.'

We arrived at Taynton Court Farm at midday, sat in the spacious kitchen, ate home-made sausage rolls and talked about welfare with James Griffiths, a tall, strikingly good-looking, dark-haired man in his early fifties. When he took over the family farm in 1990 he had just 130 milk cows on 190 acres of grassland. He now has 800 black-and-white cows which are kept indoors for most of their lives, and a recently acquired herd of Jersey cows on around 1,000 acres.

'I like to think we have a team ethos,' explained James. 'I want everybody who works here to feel involved in the decisions we take. At the last staff meeting, we talked about ways in which we could reduce our costs. I asked everybody what they considered non-negotiable. They all said the same thing: herd health and welfare.' Each of the six staff has a specific role. For example, one is in charge of dry cows during the period prior to calving; another is responsible for foot care. The welfare of the staff is also of paramount importance. They have a cast-iron guarantee of three days off every fortnight as well as annual holidays. 'The results follow the man,' said James. Every day of the year, he gets up at 4.30 a.m. to make tea for the milking staff.

When he was growing up, veterinary care was largely a question of fire-fighting, of responding to specific health problems as they arose. This has changed, thanks in part to the influence of Roger Blowey, who came up with what seemed a radical idea at the time: that farmers and farmworkers should do as much diagnosis and treatment as possible. 'This drives interest, responsibility and care within the team,' explained James. As a result, cows are treated more promptly and at a lower cost than in the past.

'We now know a huge amount about cow comfort and every-thing you need to do to provide the best levels of welfare,' explained James as we made a tour of the farm. He talked about the importance of having plenty of space for the cows to move around and lie down, and the benefits of non-slip floors, good ventilation and good bedding. The farm used to provide straw on concrete in the cubicles; now they use deep sand or sand on mattresses. This has reduced the level of mastitis, inflammation in the udder, to far below the national average.

Over the years, James has moved away from Holsteins to a black-and-white cow crossed with Swedish and Norwegian Reds. The Scandinavian breeds provide a degree of hardiness you don't find in Holsteins, which James described as being 'a knife-edge animal, needing high-quality care in certain circumstances'. His yields may not be as high as the most productive herds – he gets around 9,000 litres per year – but he believes his animals are in robust good health.

Animal welfare groups make much of the fact that dairy farm-ers generally remove calves from their mothers soon after birth. They say that this is a cruel practice. As we were walking through an airy shed housing the dry cows – the cows waiting to calve – we passed a handsome red-and-white cow with a bulging, veiny udder which had just calved, strands of rainbow-sheened after-birth still hanging from her vulva. Her calf would be removed within four hours. 'The less time they are together the better, for both cow and calf,' said James. 'There is less risk of disease, no time to form a bond, and less stress when they are parted.'

Shortly after my visit, Roger sent me various scientific papers which provided evidence that good animal welfare means better yields. Relaxed cows are more productive than stressed cows. Good welfare involves not just providing comfortable accommodation, an appropriate diet and good veterinary care, but adopting a gentle manner. Cows dislike being shouted at; they should be talked to gently, and walked slowly, for example from the fields to the milking parlour if they are grazed outdoors. 'You can always tell when cows are badly treated or frightened by the way they react to visitors,' said Roger. 'Nervous or apprehensive cows will move away when you approach them. At a farm like this, the biggest danger you face is being licked to death.'

Housed systems, where dairy cows are kept indoors most or all of their lives, are far more common on the continent than they are in the UK. For example, 90 per cent of Italian cows, 85 per cent of Greek cows and 75 per cent of Danish cows are not grazed outside. The equivalent figure for the UK is around 10 per cent. The practice is roundly condemned by organisations like World Animal Protection, which published a report when I was in Gloucestershire that claimed there are now in the region of 100 'intensive indoor dairy farms' in the UK, accounting for up to a fifth of all milk production. 'Cows in these systems never go outside, are pushed to the limits to produce more milk, and are at high risk of suffering from lameness and udder infections', said the report. All I can say is that this wasn't the case for any of the farms that I visited. The cows were kept in modern, well ventilated buildings with large cubicles furnished with the latest bedding material; all had access to the best possible veterinary care and were clearly in excellent health.

'For a while, my cows had a choice, they could either stay inside or go outside to graze in summer,' Adam Sills explained when I visited him at his Derbyshire farm. 'Most of them chose to stay inside. From my point of view, it's much easier to monitor their health, their feeding needs and their condition when they are indoors.' Sue Flowers pointed out that up in the hills life can be absolute hell for animals at certain times of year. 'People don't seem to realise that we only have about four months of summer up here and even then it's often rubbish. When we had the cows outdoors in summer, half the time they were sheltering behind stone walls because it would be blowing a gale and raining.'

The Farm Animal Welfare Council, an independent organisation which provides advice to government, concluded that there are few disadvantages to keeping cows in either large herds or housed systems, providing the stockmanship is of the highest standard. In its report, World Animal Protection, states: 'To go outdoors is a basic freedom that every cow should have.' It's certainly an appealing idea, and I personally prefer the idea of drinking milk, or eating cheese, which comes from herds that spend the summer outdoors. But then that's about me, and what I want, not about the welfare of the cows.

We know that some things are, or were, demonstrably cruel. For example, in the eighteenth century young calves that were being transported by horse and cart from Northamptonshire to Essex spent eight days on the road, with their legs tied together, sustained and sedated by a mixture of flour and gin. *Pâté de fois gras* geese are force fed. Practices such as the rearing of young veal calves in crates have been banned on welfare grounds in the UK, as has the tethering of pregnant sows. Other practices which are still allowed may be proscribed in future, the slaughter of livestock without stunning for the Halal and Kosher trade being among them.

I first met Roger Blowey on a farm walk organised by the National Farmers Union (NFU) in 2014. Its purpose was to bring farmers, vets like Roger and rural commentators like me together to discuss the issue of bovine tuberculosis (TB), a disease transmitted to cattle by badgers and vice versa. Later, when I visited him at his home, Roger talked me through the results, and the implications for dairy and beef farmers, of the government's trial badger culls in Gloucestershire. I am not going to discuss all the pros and cons of culling here – all the evidence, in my view, points to the fact that the disease in cattle will only be reduced or eradicated when good biosecurity measures, involving strict controls on cattle movements, are accompanied by the culling of diseased badgers – but it is an issue that has aroused much controversy. Bovine TB has undoubtedly caused enormous hardship to farmers and distress to cattle, over 30,000 of which tested positive for the disease in 2015 and were compulsorily slaughtered. However, the badger culls have been opposed by animal rights campaigners and some scientists on the grounds of both animal welfare and utility:

they argue that the culls are simply not working, and may even be spreading the disease.

James Griffiths was closely involved with the Gloucestershire culls, and as a result he has been subjected to considerable harassment by animal rights campaigners. His family, he explained, had frequently received threatening phone calls and his eighteen-year old daughter had recently been tailgated by a group of 'antis', who chased her down narrow country lanes late at night, flashing their lights at her. This was the bad news. The good news was that James believed that the culling of badgers on his farm had led to his herd being free of bovine TB for the first time in over ten years. This meant that he could now sell live animals from the farm, something he had been unable to do for over a decade. Early results suggested that the pilot culls had also helped to reduce the disease in cattle elsewhere.

A significant number of influential people seem to hold similar views to BBC presenter Chris Packham, who has described those involved in the badger culls as 'brutalist thugs and liars'. Most of these people, I would suggest, are not interested in the welfare of farmers or their cows; and many, it seems, care little for the welfare of badgers: death by TB – in some parts of the country, a third of badgers may have the disease – is a painful, long drawn-out affair. How strange it is that there is such outrage among animal rightists about the culling of badgers, yet scarcely a murmur of protest about the fact that many tens of thousands of foxes are shot and snared each year, mostly to protect game shooting interests, and even greater numbers of deer are culled in order to keep their populations under some form of control. Badgers, incidentally, were once part of the West Country diet and cured badger hams – the equivalent of *jambon cru* or *jamon iberico* – were often found on pub bars.

When Daniel Defoe, the author of *Robinson Crusoe* and *Moll Flanders*, journeyed on horseback around the country in the 1720s, he took a great interest in agricultural trade, whether it was corn being exported from East Anglia to Holland, turkeys being walked from Suffolk to London, cattle being trekked from Scotland to Norfolk, or cheese making its way from Gloucestershire, by cart track and river, to the capital. The Cheddar produced in the West Country was, he believed, 'the best cheese that England affords, if not, that the whole world affords'.

The last few decades have seen a great flowering of British cheese-making, and Cheddar is just one of over 600 varieties you can buy; in fact, we now produce more cheeses than the French. Among the finest I've tasted are those made by Simon Weaver, whose family has been farming in the West Country since the sixteenth century and in the Cotswolds for the last three generations.

It was a perfect day in early June, the sun genuinely warm at last, the hedgerows afloat on a white haze of cow parsley, swallows and house martens swooping above the fields, when I called in to see him. Simon is the tenant of some 900 acres at Stow-on-the-Wold – it is poor land, Cotswold brash and mostly unploughable – and 'other bits and pieces scattered around the

Simon Weaver with his low-maintenance Friesian cows.

place', as he put it, most of which he farms organically. He got into the cheese-making business after he took on the tenancy of Kirkham Farm in Upper Slaughter, a fifteen-minute drive from Stow, in 2000. At the time, the farm was in the process of going organic. 'I'm not at the goatee beard and sandals end of the organic farming spectrum,' he explained. 'I'm doing it because it makes commercial sense with the land I have.'

As soon as Kirkham Farm was certified as organic, Simon told the Organic Milk Suppliers Cooperative (OMSCo) that they could take his milk. No thanks, they replied: the supply of organic milk far exceeded demand as there had been a headlong rush into organic farming, frequently by farmers who were struggling to make ends meet using conventional farming. 'So we thought, right, we better get close to the customer and start making cheese,' recalled Simon. His wife Carol went on a cheese-making course in Germany and they began turning some of their milk into cheese.

The Kirkham Farm dairy now produces 60 tonnes of cheese a year and employs about a dozen people. Around 90 per cent is brie, but there are six other varieties as well, including a creamy Cotswold blue and a single Gloucester. The latter is made from the morning milk without its cream – hence the 'single' – combined with the evening milk with its cream. At one time, this area was full of Gloucester cows, a small, handsome breed with a mahogany sheen and a white stripe along the spine and tail, but they were replaced by more productive Dairy Shorthorns and later by Friesians and Holsteins. At the time of my visit, Simon had fourteen or fifteen Gloucester cows – it is still a rare breed but off the danger list – and he has a good story to tell: eat the cheese and you save the breed.

After we left the dairy – sparklingly hygienic and fridge-cool, the staff in white overalls and protective headgear – we wandered round to the back of the farm to look at the silage pit. Instead of taking the silage to the cows, the cows come to the silage. This self-service arrangement is very much in keeping with Simon's aim of keeping costs down. 'We're not trying to get very high yields', he explained, 'and we are close to adopting the New Zealand system.' He has small, light-footed New Zealand-type Friesians, which require less maintenance than British Friesians and thrive on a diet of grass and forage. 'If you did this to a Holstein, it would probably starve to death,' he said. The cows calve during spring and they are dried off in the winter months, which means that they produce milk in summer when the grass is at its best.

Our final port of call, before we headed for lunch – a pint of beer and a Cotswold blue salad – was Greystones Farm, near the village of Bourton-on-the-Water. The farm has always been managed in a traditional manner, without the use of agrochemicals, and its hay meadows support a rich fauna and flora, including many species of orchid. Gloucestershire Wildlife Trust bought the land a few years ago and asked Simon if he would be interested in running it as an organic mixed farm. He said that would be a good way of going bankrupt, as it was too small to run as a single unit.

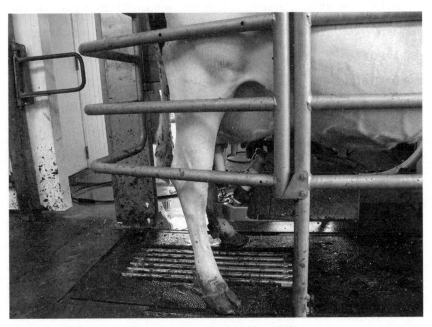

Robotic milking at Greystones Farm.

Instead, he offered to graze sixty dairy cows here and link their production to his dairy and cheese-making enterprise.

Soon after we arrived, a cow entered a robotic milking machine, recently installed at a cost of around £100,000. Rotating brushes washed her udder before a milking unit locked onto her four glistening teats. The cows at Greystones can decide not only how often they want to be milked – this is known as freedom milking – but when. In the summer months, they can choose whether to stay indoors or graze outside. While they are being milked information is gathered by a transponder on their collars about everything from their movements to their daily cudding habits and transmitted to a computer at Kirkham Farm, so the head cowman can assess their health, productivity, oestrus cycle and other matters of importance.

Before the milking machine was installed, archaeologists on the site uncovered round houses, human skeletons, grain jars, bones of hunting dogs and domestic paraphernalia. 'I like the sense that we are part of a history of land-use that stretches back to the Bronze Age,' said Simon as we watched one cow plod out of the milking machine and another take its place. This being a sunny day, the cows were all indoors, most lying on straw and gently chewing the cud. The atmosphere was one of sleepy satisfaction, such as you might find in an old-fashioned London club after a particularly good lunch. You half expected a butler to come round with port and cigars. You might get the impression that Simon was farming in an old-fashioned way. On the contrary: I think his way of farming – blending high-tech innovation with intelligent and sensitive use of resources – will become increasingly common in the future.

3   Dyed in the Wool?

The nomenclature of British sheep reads like an inventory of our countryside, the names of the breeds rolling off the tongue like rustic poetry. Badger Face Welsh Mountain, Bluefaced Leicester, Clun Forest, Devon Closewool, Exmoor Horn, Hampshire Down, Llandovery Whiteface, North Country Cheviot, Suffolk, Swaledale, Wensleydale, Welsh Hill Speckled Face: these are just a few of the ninety-odd breeds and cross-breeds developed over the past millennium. Some of the rarer ones are found in their place of origin and seldom seen elsewhere; the more popular and influential have spread their geographic wings, both within our shores and beyond.

In the fourteenth century, Edward III issued a command that the Lord Chancellor should sit on a wool bale to symbolise the importance of the wool trade in the Middle Ages – and the Woolsack is still the seat of the Speaker in the House of Lords. Sheep, wrote the agricultural historian Lord Ernle, were the sheet anchor of farming, not for their mutton or milk, but their wool, and they remained the chief source of trading profit for English farmers up to the Industrial Revolution. In the eighteenth century, travellers like Daniel Defoe were astounded by the number of sheep they saw. When Defoe visited Dorchester, he was told that 600,000 sheep were fed within six miles of the town. 'I do not affirm it to be true,' he wrote, 'but when I viewed the country round, I confess I could not but incline to believe it.'

Today, sheep farmers make little from their wool. They pay approximately 90p to £1.10 for the shearing, or clipping, of each sheep, unless they do it themselves, and in return they get between 50p and £3.00 per fleece, depending on its quality. It is the trade in meat, over a third of which is exported, which determines the shape of the modern sheep industry. Approximately 40 per cent of the breeding flock is based in the uplands. While some hill farmers rely entirely on sheep, many also run herds of beef cattle which are well suited to the terrain and the weather. The Mediterranean has its agricultural trilogy of wheat, olives and vines, first imposed on the region by the Romans; in the Pennines, and elsewhere in the uplands, we have the hardy trilogy of sheep, beef and grouse.

Two years running, just after Midsummer's day, I made my way up Nidderdale to the village of Middlesmoor, which gives its name to an estate of some 10,000 acres. I chose to come here for two reasons. First, this part of Yorkshire is a stronghold of Swaledale sheep, which have been familiar to me since my childhood. I was also advised by farming friends in neighbouring Wensleydale that Middlesmoor's owner was a particularly enlightened member of the landowning class. On my first visit to Middlesmoor, I stayed at the Crown Hotel, a handsome stone building with a congenial landlord, wholesome food and fine views across the steep valley; the following year, I parked my motorhome in the field behind the Crown.

This is fine walking country and one morning I headed up the drover's road above the village in the hour before dawn. The only sound came from the scuffing of my boots on the rocky track and the occasional bleating of a lamb. Light gradually seeped into the eastern sky and by the time I reached the highest point the distant hills had swum out of the darkness. A lapwing shrieked above my head and before the sun had fully risen curlews were calling and I heard the *go-bak go-bak* call of grouse in the heather. A straggle of black-faced ewes briefly stopped nibbling the dewy herbage and lifted their heads to gaze disinterestedly in my direction.

I had walked up here before, once in heavy rain and on another occasion under blue skies in the company of a flock of recently shorn sheep. What struck me most, then and now, was the sheer density of human habitation in Upper Nidderdale. You would think buildings would be few and far between in this world of rough pasture, heather and rocky outcrops, but the opposite is true. Everywhere you look, there are farmhouses and field barns, and the moors are pocked with old quarries and mine shafts and water channels. One of the farmers I met here showed me a lintel with the date 1655: 'Can you imagine people living up here in those days, all because of the value of wool?' he asked. In fact, those seventeenth-century sheep farmers were relative newcomers. Some families in the dale can trace their ancestry back to the early Middle Ages.

John Rayner was watching two men – I think they were his sons – running sheep through a foot dip when I called in to see him at his farm overlooking Gouthwaite Reservoir, a few miles

John Rayner beside one of his old field barns.

downstream of Middlesmoor. When they had finished with the sheep, he led me through an old cattle byre to the farmhouse, a handsome stone-built affair with mullioned windows. It was once part of a fine manor house, Gouthwaite Hall, and it had been taken down, stone by stone, to be rebuilt here, after the waters of the new reservoir began lapping through the front door in the late 1890s.

We looked at a photograph taken in 1910 of the annual sheep washing – sheep used to be washed before they were clipped – and at newspaper cuttings about John's family, which he could trace back to 1215. Before the Dissolution of the Monasteries in the sixteenth century, the Rayners supplied butter and eggs to the monks at Fountains Abbey, one of the greatest of all Cistercian monasteries. They have been farming in Nidderdale ever since. Like most farmers in the dale, they have beef cattle as well as sheep.

'I were born in t' thirties,' said John curtly when I asked how old he was. A tall, rangy, sharp-featured man with mutton chop sideburns, pale blue eyes and a shock of white hair, he could have stepped straight out of the pages of *Wuthering Heights*. After we had looked at more photographs and an old notebook recording livestock sales, we set off up the hillside at a brisk pace. When we came to a small herd of suckler cows, Limousins crossed with other beef breeds, one of the cows began stamping her hooves and swaying her head. 'Just look at way she's looking at you,' said John. 'There's summat wrong in t'ead wi' Limys.' He explained that he had recently been obliged to remove the cows from a field with a footpath. 'They kept tossin' women in long dresses over t'edge.' Apparently, they would charge halfway across a field for the satisfaction of doing so.

We clambered up a track beside a row of mature hazelnut trees, inspected a mole trap – some putrid corpses were rotting on a stone gatepost, a warning to others perhaps, or just proof of success – then came to a fine slate-roofed field barn where cattle were once housed in winter. John's father used to have a herd of Dairy Shorthorns and John showed me a dusty old milk churn with an oval plaque inscribed: 'M. Rayner Gowthwaite Pateley Bridge'. Every morning, his father used to take the churn by horse and trap to Pately Bridge, a nearby market town, to put it on the train for Leeds.

In the upper level of the barn, where the Rayners used to store hay, there were some ancient bracken bales – it was cheaper to cut and bale bracken rather than buy straw bedding for the cattle – and a grass cutter, bought in Ripon in 1891 by John's great-grandfather for £7. When he was young, they used to pull the grass cutter with horses; later they converted it for use with a tractor. 'I suppose I should put it in a museum,' he said, 'but in winter I like coming up and greasing it every now and then.' After we left the barn we climbed further up the hill to a 30-acre rabbit warren, dating back several centuries, entirely enclosed by a stone wall. Away to the right, John pointed to a jumble of large stones, the remains of a Bronze Age settlement.

The brief time I spent at the farm – one of many thousands where sheep have been tended for century after century – reminded me of the obvious, but easily forgotten, truth: that we pass fleetingly through this world like seeds from a dandelion clock blown across a meadow. This, in turn, reminded me of something Stephen Ramsden, the owner of the Middlesmoor Estate, had said when I first met him: 'I'm just a custodian for this landscape, passing it on to the next generation.' It is the sort of thing many landowners say. In Stephen's case, he really meant it.

I first met Stephen in 2015. By the time I arrived at Middlesmoor, the rain that had followed me across the moors had cleared and white clouds like well-washed sheep were skidding across the high country. There was nobody at the Crown, so I sent a text to Stephen telling him that I had arrived. Reception was poor and when he phoned back he had to shout several times before I got his message: I would find him in a hayfield behind the hotel. He finished raking the grass into rows and we shook hands when he got down from the tractor. He was everything you don't expect of a man who owns ten farms, a large grouse moor and a good chunk of the village, including the Crown. He was dressed like a farmworker, in oily overalls, and he had the hands of a man who does plenty of manual work.

We sat at a wooden table outside the pub, with fine views past the church tower, and Stephen told me how his family had come to Nidderdale. His great-grandfather, who was in the colliery business, bought the Middlesmoor Estate in 1919 and put it in his son's name. Stephen's grandfather survived the First World War, but volunteered for the next war and was killed in Norway. His son, Stephen's father, was seven at the time, so the estate was run by agents in Cheshire, where the family lived, up to 1957. From then until 1989, it was managed by Stephen's father, but he had a stroke at the age of fifty-six and never recovered his speech.

'Before I took over the estate in 1989 – I was twenty-eight at the time – we had always been absentee landlords because the family's main interest was grouse shooting,' recalled Stephen. Within a couple of years of his arrival, he had married the daughter of a tenant farmer and taken over a gamekeeper's house. He now farms around 1,000 acres himself, three-quarters of which is grouse moor. 'Best thing about being an estate owner is that it's a committee of one – and that's me. As a landlord, I'm very hands-on. Back in 1919, there were seventeen farms on the estate. Now we have ten and I'm determined to keep them all going.'

10,000 acres sounds like a large estate: imagine a block of land 5 miles long by 3 miles wide. However, size doesn't equate with profitability – a 50-acre farm growing vegetables in the rich soils of Cambridgeshire would be far more profitable than the whole of the Middlesmoor Estate – although it is true that much of the countryside remains in the hands of large, and for the most part, wealthy landowners. Stephen Ramsden is a minnow in this particular pond. All the same, the way he manages the estate has a profound influence on the social and economic welfare of the upper dale.

At the beginning of the twentieth century, 120 people lived in Middlesmoor; now there are just forty. Around a third of the houses are owned by Stephen, a third are owner-occupied and a third are holiday cottages. 'But we are doing much better than many other places in the Yorkshire Dales,' he said with satisfaction. Over two-thirds of the dwellings in Arkengarthdale and half in Kettlewell, in Wharfedale, are second homes. 'If people have been born and bred here, I want to give them the opportunity to stay, that's why I am happy to offer new tenancies to the children of my tenants – providing they look after their farms well.'

Soon after he took over, he began a programme of modernisation as no money had been spent on the farmhouses for some forty years. He also did his best to weatherproof, at considerable expense, the twenty-eight field barns on the estate, the vast majority of which no longer served any agricultural purpose. In the past, cattle were wintered in the lower floor and hay stored above.

'Why didn't you just let them fall to bits if they weren't being used?' I asked.

'Oh, I couldn't have,' he replied indignantly. 'They're part of the fabric of the landscape, just like the stone walls.' He has 40 miles of dry stone walls on the estate, as well as 12 miles of public rights of way, all of which he is obliged to look after.

Early one morning Stephen picked me up after breakfast and we juddered up a rough track. It was one of those wonderful days you sometimes get in the Pennines, the sky blue and almost cloudless with a warm breeze rustling the long grass and meadow pipits and skylarks trilling above our heads. After about 4 miles we spotted several farmers in the distance, some on quad bikes, others on foot with their sheepdogs. We clambered across the heather and Stephen introduced me to his father-in-law, Alan Firth, before heading back down the dale.

A friendly, loquacious, well-nourished character with a round face and spectacles, Alan explained that there were about a dozen farmers and helpers – some below us, others way up the hill still – driving the sheep towards two lots of ancient stone washing pens. This was his fifty-fourth year of gathering sheep and the very thought stimulated a tumble of memories.

It was so cold up here one winter in the 1980s that he found sheep standing immobile, like wooden tables, their heads and necks iced to their spines. 'We cracked the ice on the back of their necks

Alan Firth at his fifty-fourth sheep gathering.

with an iron bar', he recalled, 'and then they would just scuff away the snow with their feet and start nibbling at the grass. You'll not find a hardier sheep than Swaledales.' On another occasion he and another shepherd led a flock of sheep back through the snow to a farm near Middlesmoor village in single file, enticing them on by dragging a ball of hay in front of them. They all survived, but there were other occasions when they were less fortunate. 'Four or five years ago, we came up here in the middle of winter. Lots of sheep were stuck in snow drifts, but there was nothing we could do about it, so they died where they stood.'

As the sheep were being driven down the hill I felt as though I was witnessing a scene from much earlier times, but that was a romantic illusion. There had been many changes, mostly for the better, since Alan first came to Middlesmoor in 1962. All the farmers had quad bikes now, which made it easier to get around the fields and moors. In the old days, you could be snowed in for days or even weeks. Now farmers had large tractors and they could swiftly clear the snow. However, there was less shepherding than there used to be. 'The young, they all have second jobs, contracting and the like, so they don't have so much time to spend with the sheep,' said Alan. 'We used to come back the day after the gathering to get the stragglers. Now, any sheep we don't get today will be left here and they won't get clipped.'

After an hour or so the sheep were corralled into one set of washing pens while another lot was driven into the pens a few hundred yards away. When Alan and I reached the lower set of pens we joined a group of farmers – five men and a blonde-haired farmer's daughter – who were eating their lunchtime sandwiches. When they finished, they began separating the sheep into different pens, so that each could set off for home with his own flock. Sorting the sheep was none of the sheepdogs' business, but they frequently jumped into the pens and their indiscipline added to the theatre of the occasion. I headed back to Middlesmoor behind one flock of sheep and ahead of another. Eventually, I caught up with the first flock. Mike Pyete, part-time farmer, part-time electrician, had been hired for the day and he was bringing up the rear of Alan Firth's flock on his quad bike, a lame sheep clamped between his legs. I remarked that it was an impressive sight, looking back towards the hills and seeing so many sheep heading towards Middlesmoor.

Mike Pyete drives Alan Firth's flock of sheep back to Middlesmoor.

'This is nowt,' he replied. He produced a phone and flicked through photographs until he found what he wanted: several shots, taken from his quad bike, of a 1,000-strong flock of sheep that he had single-handedly taken back to the moors after clipping a few years ago. 'They stretched so far into the distance you could hardly see the end of them,' he said.

He wanted to know whether I'd seen the film *Addicted to Sheep*, a documentary about a Swaledale sheep-farming family in Teesdale. I said I had, and that I had loved it. It captured the hardiness, good sense and humour of the shepherding world. 'They're friends of mine, and I give them a good tease about the film when

I see them,' he said with a grin. I asked whether he had read James Rebanks' *The Shepherd's Life*. He nodded: 'Yes, it were quite good.' High praise in this part of the world.

The Shepherd's Life is an account of a sheep farmer's life in the Lake District, written by somebody whose family has farmed there for 600 years. Rebanks describes his attachment to landscape and community, the buying and selling of stock, the seasonal routines of sheep farming, the tension between locals and outsiders, and traumas great and small, from the 2001 foot-and-mouth epidemic to the losses caused by sheep-worrying dogs. Sheep farming, he reflected after he left school, was a tough business: 'Our sheep made the same price at market as they had done twenty years earlier. We kept more and more and made less and less money… Our buildings were thirty or forty years old and slowly falling apart… Farming was changing too, with a raft of new regulations that would cost a fortune to abide by on an old farm like ours.' Yet Rebanks was determined to make a go of sheep farming. Not only has he succeeded in these terms, he has provided a window onto a world that most townspeople know little about. In so far as there are any farmer celebrities, Rebanks has become one, with around 100,000 followers of his Twitter account.

If you want to get an insight into the complexity of the world of sheep, and the variety and virtues of different breeds, you should go to a sheep fair or an agricultural show. Before my first visit to Nidderdale, I attended Sheep South West 2015, the biennial fair organised by the National Sheep Association, on that occasion held at a farm near North Tawton in Devon. There were several thousand people there and I was probably the only one who wasn't a sheep farmer or closely associated with farming. In one marquee there were some forty breeds of sheep on display, with three or four of each breed enclosed in a small pen on a bed of straw.

While I was wandering through the marquee, I heard a voice which sounded familiar; it wasn't just the tone of the voice and the intonation that caught my attention but the accent, which was from the northern Dales. I took a good look at the tall, portly, rosy-cheeked individual standing beside the Swaledale pen and eventually realised who it was. I hadn't seen John Stephenson, a farm student on the Clifton Estate in Masham the year after I was there, since the early 1970s. John was now the secretary of the Swaledale Sheep Breeders' Association and he suggested I look him up next time I was in Yorkshire.

When we met for lunch many months later at the Rokeby Inn, on the A66 near Richmond, John gave me a copy of the latest Swaledale flock book. Among other things, the annual flock books enable farmers to trace the ancestry of a sheep back to earlier times, although he admitted that the judging of Swaledales had much to do with subjective ideals of beauty defined by the early breeders. The flock book goes into great detail about what a good Swaledale sheep should look like: horns should be set low, round and rather wide; ears should be grey or silver and of medium length; the tail should be thick, long and woolly; front and hind legs should be grey or mottled in colour at the front and black at the back with a silver tip on the hock.

I told John that one of the Nidderdale farmers I had met – he kept Scottish Blackface sheep, rather than Swaledales – had a message for him. 'He says your Swaledale sheep shows are just Cruft's for sheep, and that the Swaledale Sheep Breeders' Association is obsessed by the aesthetic appearance of the breed, rather than the attributes that make a good commercial animal.'

John's cheerful, ruddy face creased up. 'Oh, well, I might as well leave now then,' he replied, affecting disapproval for any attempt to disparage the breed.

Even if Swaledale farmers don't show their sheep, they all seem to take pride in their appearance. Their main aim, however, is to make money. Most Swaledale ewes are crossed with Bluefaced Leicester tups to produce a northern Mule. Most of the male Mules – the wethers, as they are known – are castrated and fattened for the meat market. The females, or gimmer ewes, generally end up on lowland farms, where they are put to Suffolk, Texel or Charollais tups. It is all about hybrid vigour and getting the best general-purpose lamb for the meat market. While the majority of Swaledale sheep farmers try to produce good quality Mules which will fetch a reasonable price in the market, a small number spend considerable time and money trying to get into what John described as the top section of Swaledale sheep. 'The dream is to produce a tup worth £40,000 or more, like the Ewbanks in Middlesmoor. Have you met them?'

I called in to see Mark Ewbank on the way back from the sheep gathering. Mark was clipping sheep in a barn, watched by two young children and helped by his father, Spencer, a sprightly, quietly spoken seventy-nine-year-old who lived down the dale but came up to help at the farm every day. As we walked across to the farmhouse, Spencer explained that his father, who had been wounded at the Battle of the Somme – we had just had the 100th anniversary – took on the tenancy of Intake Farm during the 1930s depression. At that time, British farmers were unable to compete with the cheap cereals, beef and other agricultural goods that flooded into the country from the Americas and elsewhere. Farmers couldn't even sell their wool and many tenant farms went unlet.

Spencer took over the farm from his father in 1965 and retired in 2007, handing over to Mark, who was also gently spoken, but built on a larger scale, like a rugby player, with a wispy beard and moustache. It was Mark who decided to bring about improvements in their Swaledale flock, his first measure involving the purchase of a high-quality tup for £5,000. It was well worth it, he said: the tup fathered over 300 ewe lambs and he sold one of its male lambs for £4,600. The walls around the kitchen table were hung with photographs of their best stock, including a tup which had sold for £52,000 at Hawes, a market town in Wensleydale. Mark agreed that looks and appearance played a major role in determining its value.

'I'm surprised you didn't faint when you heard the price it went up to,' I said.

'No, I didn't faint,' he replied.

'He just turned white,' said Spencer.

'But was it really worth £52,000?' I asked.

'Well,' replied Mark. 'The first lot of its progeny earned the buyer £150,000. If you look at it like that, it was.'

A few months later, I went to the autumn sheep fair at Wilton, near Salisbury in Hampshire. Several thousand northern Mule ewes were on sale – many would have come from Yorkshire – as well as fifty or sixty Texel rams, most of which were fetching around £300 each. Texels are well-muscled sheep of Dutch origin

Crufts for sheep? Awaiting inspection at a Hampshire fair.

which blend well with other breeds to produce high-quality lean carcasses. I fell into conversation with one of the Texel farmers and asked whether the tups were judged on their beauty, like Swaledales, as well as their utility.

'Swaledale farmers are living in fairyland when it comes to some of the prices they pay for tups – but good for them,' he replied. The answer to my question was no: the buyers here were looking for good conformation, which wasn't the same as beauty. 'What I want in a Texel is a long straight back, and I want to see it's well ribbed-up and has a good back end. It's like the way you look at beef breeds. I was just watching a young man back there feeling a Texel's testicles. That tells you a lot about what matters, that and their genetic history.'

If you consult TripAdvisor, you will see that the Crown Hotel in Middlesmoor gets some tremendous reviews and a fair number of shockers. 'Why this man is working in the hospitality industry is beyond me!' wrote one dissatisfied reviewer. This man was the admirable Malcolm Whitaker, a gamekeeper turned publican who has the figure and complexion of a man who enjoys his food and drink. On my first night at the Crown I ordered a ginger beer with the meal. A theatrical eyebrow was raised behind the bar, so I told Malcolm that I was on a new regime: four or five nights a week off proper drink, the others on. The following night, I asked for a glass of red wine with my lamb shank. Malcolm filled a large glass to the brim, then slapped the bottle down on the table, declaring that what remained, a good four fingers of French *vin de table*, was just an embarrassment and the sooner I drank it the better. By the time I got to the treacle tart, Malcolm had abandoned his post at the bar and settled down for a meal with some of the locals, who were accompanied by a menagerie of gundogs and terriers. Snatches of conversation and bursts of laughter drifted into the side room where I was eating: '... and t' bloody thing was... aye, but 'e were a good jockey... she spent 'er time cheffin' and ridin' 'er 'oss...'

On another occasion I was joined at dinner by the only other guest, a recently retired chap from Suffolk who had come for a map-reading course in an open-top MG Midget, accompanied by a fox terrier which he addressed as 'sweetheart'. He was itching for a decent conversation, but I disappointed him. When I told him why I was here, he reeled off a list of writers on country matters who he had either read or heard at literary festivals, including Richard Mabey, George Monbiot, Robert Macfarlane and Helen Macdonald.

I told him that I had little appetite at present for nature writing, especially when it was a camouflage for a misery memoir, or for environmental polemics. I preferred poetry to poetic prose, or more practical works. My current bedtime reading was *Profitable Sheep Farming*, the 1983 edition of which I had recently bought in a second-hand bookshop on Exmoor, and Owen Sheers' *A Poet's Guide to Britain*. This featured many of my favourite poets, including Gerard Manley Hopkins, R. S. Thomas, Seamus Heaney and

Ted Hughes. The poems by Thomas, who was a vicar in the Welsh hills, give an unflattering yet vivid portrait of impoverished and often lonely sheep farmers. Seamus Heaney's poems about toiling on the land come from childhood experience. And Sylvia Plath, Ted Hughes' first wife, is marvellous on sheep: *They stand about in grandmotherly disguise,/ All wig curls and yellow teeth/ And hard, marbly baas.*

Plath wrote these lines in 1961, a year before the introduction of the EU Common Agricultural Policy (CAP), which has had a profound effect on our landscape and farming systems. In the 1930s, when Mark Ewbank's grandfather took over Intake Farm, there were about 25 million sheep in the UK. This figure remained fairly constant until the late 1980s, when various sheep-related subsidies were introduced, leading to an explosion in numbers – the sheep population reached 45 million in 1999 – and serious overgrazing in many parts of the uplands. As a result of the subsidy system many parts of the uplands have been sheep-wrecked, to use the expression coined by the environmental journalist George Monbiot. The foot-and-mouth epidemic of 2001 led to the slaughter of 5.5 million sheep, but numbers are now back to around 33 million.

Stephen Ramsden didn't think that overgrazing was a problem on his estate. 'I allocate the annual grazing licences for my tenants, and that means I control the number of sheep in the high country,' he explained. 'We've always had to think about grouse, so it has been in our interests to make sure there aren't too many sheep.' If there was no grouse shooting, stocking densities would probably be much higher and this would have a damaging effect on the vegetation and wildlife.

Grouse shooting, incidentally, is big business in Nidderdale and the sport fits well with sheep farming. I asked Stephen what would happen if driven grouse shooting were banned, as some environmental campaigners wished, largely because they believe gamekeepers on grouse moors are killing birds of prey. 'It would be a disaster. Wages for the beaters and loaders amount to around £1,500 a day on this estate – that works out at £50,000 a season if we have a good year. That's a large amount of money going straight into the pockets of local people.' If there was no grouse shooting, he also doubted whether the Crown Hotel would survive.

My first visit to Nidderdale took place a year before the referendum on EU membership. All the same, the subject of agricultural subsidies cropped up in many conversations. As far as these upland areas are concerned, there are two main types of subsidy. The first, for which all farms are eligible, is the area-based Basic Payment Scheme, or the Single Farm Payment, as most farmers still call it. Around one-third of this comes under the heading of 'greening' and requires farmers to provide environmental benefits in return for these direct payments. The second tranche of subsidies supports environmental stewardship schemes and most of the farmers I spoke to in Nidderdale were receiving payments – as much as

£30,000 a year in the case of some – for reducing stocking densities and farming the land more extensively.

One of the more thought-provoking encounters I had during my time in Nidderdale was with Andrew Hattan, a handsome, stockily built man in his mid-forties. He was unusual for a hill farmer in that he had studied agriculture at university, gained a PhD and worked on a range of lowland farms before taking over a tenant farm at Middlesmoor. He came to find me at the Crown one afternoon. It had rained heavily in the morning and the countryside glistened under a bright sun as we made our way on foot through a series of flower-spattered meadows that descended to the stream below Middlesmoor.

When we reached the thin seam of woodland in the valley bottom, Andrew said he intended to clear the bracken and plant native trees like alder, ash and oak. This would be good for the wildlife along the river. As we began the ascent to his farmhouse, through meadows tinged yellow with buttercups and hay rattle, he explained that he had signed a stewardship scheme with Natural England which will provide him with a significant sum of money every year for ten years. One of the main aims is to restore five old hay meadows. He was particularly pleased that several pairs of redshank, a wading bird which had been absent when he took over the farm eight years ago, had nested in the meadows this spring.

'At the moment, we simply couldn't survive without the Single Farm Payment and the stewardship money – and for that matter, the money my wife earns working part-time as a dentist. But sooner or later at least some of the subsidies are going to come to an end. What I'm trying to do over the next ten years is make sure that we will make enough profit from farming enterprises to stand on our own two feet.'

Like most farmers in Nidderdale, Andrew had a small herd of cattle, as well as a much larger flock of sheep. He had recently sold his Belted Galloway beef cattle and bought some twenty Northern Dairy Shorthorn cows and a Shorthorn bull. He was running them as a suckler herd – the spring-born calves stay with their mothers till they are weaned in the autumn – but he planned to start milking the cows and manufacturing cheese. This way, he believed, he would make more money than he could ever make from a beef herd. 'In places like this, we need to work to our strengths. We are never going to be big food producers, but we have a fantastic story to tell. We need to use our imagination and add value. What I'll be selling is cheese that comes from grass-fed Shorthorns that graze on meadows that are rich in flowers and other wildlife.'

After tea with his family we walked up to the fields behind the farmhouse to look at the Shorthorns. Andrew returned, once again, to the subject of subsidies, explaining that a farm business survey in the dale had looked at the accounts of five farms over two different periods. These showed that they were receiving subsidies which were two and a half times greater than the amount they made from the sale of their livestock.

I said I wasn't surprised. I had recently been to a seminar in London organised by the Department for Environment, Food & Rural Affairs (Defra) on risk management and farming. One of the speakers had analysed data for cereal producers from the latest Farm Business Survey, and this revealed that 43 per cent of their income came from the sale of farm produce, 39 per cent came from the Basic Payment Scheme and 12 per cent from agri-environmental schemes. In other words, over half the money earned by cereal farmers – and this included many of the largest landowners in the country on the best land – came from the public purse. So it wasn't only smaller operators who were receiving generous subsidies, but major food producers using the most modern farming techniques.

In 2016, I visited Nidderdale just a week or so after the Brexit vote and there was much consternation among the farmers about the implications for areas like this. A cartoon in the *Spectator*, published shortly before the referendum, summed up the mood. A tough-looking urban dog with a malevolent grin tells two sheep: 'You'll lose all your subsidies if we leave Europe.' The sheep farmer turns to the dog owner and says: 'Your bloody dog is worrying my sheep.'

As I write this, I can see a farming magazine on my desk. The front page headline, written in response to the Chancellor of the Exchequer's announcement that the current system of agricultural support will be retained until 2020, says: 'FOUR MORE YEARS – BUT THEN WHAT?' The article voices a sense of grievance that farmers may be deprived of the current level of support after 2020. Whatever the rights and wrongs of the argument, I have to say that its tone reminds me of a spoiled teenager, affronted that his or her rights – to iPods or clubbing in Ibiza or respect – are under threat.

This brings me back to Rebanks. *The Shepherd's Life* has been deservedly praised. Yet there is one significant omission in the book. Nowhere does he mention that he and his sheep-farming neighbours in the Lake District have survived largely thanks to subsidies. Indeed, it was the post-war support system that enabled hill farmers to become full-time farmers. In the past, their fore-bears had been part-time farmers, dependent for most of their income on activities such as mining and quarrying.

Andrew Hattan was well aware of the fact that it was the taxpaying public whose largesse had enabled him to survive. 'I'm very grateful for the subsidies,' he said when we first met. 'I pinch myself when I wake up every morning and think how lucky I am to be working on this farm and living here.' When I saw him again at the sheep gathering he asked: 'Do you think people in Leeds or Bradford will want to pay subsidies to places like this in future? Mightn't they prefer their money to go to the steelworks in Wales or Redcar, or the NHS?'

Some might. However, I believe that a significant body of the taxpaying public cherishes places like the Yorkshire Dales and the Lake District. These are man-made landscapes, created over the centuries by the ancestors of today's farmers. If they go out of business, which many would without support, large areas of land would be abandoned and presumably turn to scrub. The case for continuing to provide support for farmers in areas like Nidderdale is strong, but rather than providing area-based payments, which reward farmers of all shapes and sizes, regardless of their wealth or poverty, I would suggest that the payments should be for

Andrew Hattan, with Alan Firth behind, sorting sheep.

services rendered. This includes adopting measures to protect wildlife, such as those which are encouraged by existing steward-ship schemes; looking after landscape features like stone walls; and providing a range of other public goods such as clean water or – in the case of moorland owners – carbon storage in the peat.

But I think Andrew Hattan is right when he says that hill farmers should be making greater efforts to establish commercial enterprises, for example by adding value to what they produce. I had a similar conversation when I met John Stephenson for lunch. He also said he believed that farmers needed to pay much greater attention to market forces. 'There are still a lot of people in the Dales who don't know what's happening in the wider world, who don't really understand what the public wants,' he said. 'If you are selling doors or windows, you'd make sure that you produce exactly what people want – but most sheep farmers don't think like that.'

One move in the right direction began some six or seven years ago when John met the meat manager for Marks & Spencer at the Royal Ulster Show. The manager asked John whether he could get enough farmers together to provide a guaranteed supply of Swaledale lambs between mid-February and mid-April. During the first year, John organised the supply of some 250 lambs a week; now it is up to 500 lambs a week. 'It's not earth-shattering,' he said, 'but it does mean that specially branded Swaledale lamb meat is getting into M&S during the springtime, and that's been good for the farmers who supply the lambs.' It's all about eating the view, he explained: linking Swaledale sheep to a beautiful upland landscape. Eat our lamb and you're helping to protect the Dales.

Not that the scheme has always run smoothly. At one point, M&S said they could only take a reduced number of lambs for a period of a few weeks. This meant that some of the forty-odd farmers who supplied lambs would either have to forego sales for this period, or that everyone would have to sell fewer than expected. 'You wouldn't believe the grief I was given by people down the phone, all the effing and blinding, when I explained what was happening,' recalled John, shaking his head. 'And I thought these people were friends of mine!' He sounded genuinely hurt. 'But then farmers can be like that. They often seem to just care about themselves.'

So what of the future? 'There will eventually have to be less reliance on the 'brown envelope' — the subsidies — but it's hard to see how that'll work out. When and if that happens, we'll find out who the best farmers are,' he said. I left the pub thinking how fortunate the Association was to have John as secretary: diligent, honest, passionate about the Swaledale breed, but objective enough to recognise that the stubbornness of farmers, and their independent spirit, doesn't always work in their favour.

4   The Fruits of Success

Although post-war food rationing had almost come to an end by the time I was born – restrictions on the purchase of meat were finally lifted in 1954 – my parents' garden in Yorkshire had been shaped by more austere times. There were flowerbeds in front of the house, but most of the back garden was devoted to fruit and vegetables. There were two long rows of raspberry canes, a dozen or more blackcurrant, redcurrant and gooseberry bushes, a good supply of rhubarb in a shady corner near the potting shed, and strawberries which attracted shrews, hedgehogs and small children. We had a small, unkempt orchard too.

On my travels around the country I focused on two fruits, inspired in part by memories of my childhood garden. It is the apples and strawberries I remember best: the former as trees to climb and be dazzled by at blossom time; the latter as the most longed-for crop at the height of summer. But there was another good reason for choosing apples and strawberries. Although the amount of land dedicated to fruit in the UK has fallen by a third since 1985 – from 45,000 hectares to 30,000 hectares – there have been two great success stories: the first involves cider, the second soft fruit, and especially strawberries.

In 2005, cider represented just 1.2 per cent of alcohol sales; by 2012, this figure had risen to 9.4 per cent. A drink that was once a great favourite of middle-aged men and down-and-outs has become popular with young people, and especially women under the age of thirty. Cider doesn't count as one of your five servings of fruit and vegetables a day, key to a healthy diet, but the consumption of soft fruit does and this is one of the reasons why sales of berries have more than doubled over the past decade. By 2016, berry sales amounted to £1.1 billion, half of which went on strawberries.

Most of the stories told in this book are about primary producers; about the men and women who are selling lambs or pigs or cattle or vegetables to businesses who process and package them for sale to the public. There have been some exceptions, like the Belchers in Leicestershire, whose enterprise I described in the first chapter, and Simon Weaver, the cheesemaker from Gloucestershire, but this is largely a book about farmers. The fruit industry, however, provided me with an opportunity to explore the whole process of production, from apples to cider, from strawberries to jam, within the same businesses. In short, it is about farmers as manufacturers.

I decided to spend time with two very different breeds of cider maker: Hunt's Cider, a relatively small family business dating back two centuries, and Thatchers Cider, one of our largest and most successful family-owned drinks companies. However, they were not the first cider businesses I came across. By the time I saw them I had already begun to develop a taste for cider, and the culture which surrounds it, while visiting dairy farmers in Somerset. One of them suggested I drop in at Land's End Farm, in the village of Mudgley.

The farm was at the end of a narrow lane surrounded by high hedges, sloping orchards and small fields with beef cattle and sheep. There was a sense of weathered antiquity about the farm buildings, and the three gents who were there when I arrived at 11 o'clock in the morning – they were each holding a glass of cider – looked as much a part of the scenery as the cattle grazing on the Somerset Levels below the farm. One of them was a farmworker. Another seemed to be on a social visit. The third, and the most forthcoming of the three, introduced himself as John Hallam and immediately explained that he had come to buy a consignment of cider to sell to pubs and restaurants. He had long grey hair and a grey beard and a weathered complexion. The owner of the farm, Roger Wilkins, had gone out but John said he'd be happy to show me round.

'This place hasn't changed much in the thirty or more years I've been coming here,' said John approvingly. At the business end of the barn there was a hydraulic cider press, a washing pit and a small farm shop with flagons of cider, chutneys, vegetables and other local produce. On the way to the lower level of the barn, which was furnished with battered old sofas, we passed four large wooden barrels with taps from which the cider is drawn. Decades, or possible centuries, of drips had carved deep runnels into the floor. There were cobwebs everywhere and the place had a vinegary smell about it, like a man who had drunk long and hard.

After John had shown me round the barn he talked about the trends for ever larger companies to buy up the small, of the importance of drinking cider with a good cheese, preferably Cheddar or Stilton, and of the ghastliness of fruit-flavoured ciders. 'You shouldn't even call them ciders,' he said, shaking his beard in disapproval. 'They are not ciders. They are just alcopops in a box, and most of them aren't even made with proper fruit, but flavourings!'

He then told me a curious story. 'I once turned up around 8 o'clock on a November night and there were four Japanese here. One of them had a bit of English, but the others couldn't speak a word, and they were standing about, tasting the cider. After a while, this taxi driver came in and explained that he wanted to see what they were up to.' The Japanese had landed at Heathrow – well over two hours' drive away – and told the driver to bring them here. They didn't have the correct address, just enough information to track the farm down. 'Yes, this place is known all over the world,' mused John. It is not that the cider is particularly unusual; it is more to do with the way in which the relationship of the man to the land and the orchards to the landscape has created something special. Coming here, for many, is an act of homage. Cider is that sort of a drink.

Cider was widely drunk in Britain when the Romans arrived and although its popularity has waxed and waned it has always been strongly associated with the West Country. In Somerset alone, there were over 9,700 hectares of cider orchards in 1894, which is a third more than there are in the whole country today. It is the only drink I can think of which has its own soundtrack, with the Worzels (greatest hits: *I am a Cider Drinker* and *The Combine Harvester*) being the leading exponents of Scrumpy and Western. And in my imagination it is cider, not beer or gin or any other drink, that lubricates that rustic rite of passage, kissing outdoors for the first time.

'Never to be forgotten, that first long secret drink of golden fire, juice of those valleys and of that time, wine of wild orchards, of russet summer, of plump red apples, and Rosie's burning cheeks,' enthused Laurie Lee in *Cider with Rosie*. 'I put down the jar with a gulp and a gasp. Then I turned to look at Rosie. She was yellow and dusty with buttercups and seemed to be purring in the gloom; her hair was rich as a wild bee's nest and her eyes were full of stings.' You can imagine the rest: this is what cider, and a beautiful girl, does to a teenage boy.

Long anticipated, much doubted, the British summer always comes as a surprise when it finally arrives. A great lushness had settled over the southern counties when my wife Sandie and I headed west in early June. In Wiltshire, there was a hint of yellow on the barley, the first sign of ripening. In Somerset the hedgerows looked like vertical salad bars, the green leaves of nettle, bryony, bramble and wild mint garnished with the flowers of red campion, white stitchwort and purple foxglove. Along the south coast of Devon oak and ash were in fresh foliage and the apple orchards, now in full blossom, looked like fluffy white soufflés from a distance.

When we arrived at Broadleigh Farm, home of Hunt's Cider, after a tortuous last few miles down narrow lanes with high hedges, we were shown where to park our motorhome by Roger Hunt, an avuncular character in his sixties with a sun-tanned face and a melodious accent that suggested that time moves at a gentle pace in this part of the world. I imagine the Hunts probably make as much money from their caravan site as they do from their orchards, and Roger explained that there was a symbiotic relationship between the two. Tourists come here because they love staying on farmland just a few miles from the sea; they are also keen drinkers of the farm cider.

Richard Hunt with the new cider barn behind.

Before dinner we drank a bottle of Hunt's (their apostrophe, not mine) Sparkling Heritage Cider, which had a tart zing to it; then we moved on to the medium-dry Devonshire Farmhouse Cider as an accompaniment to a duck and apricot casserole. A skilled cider maker, or accomplished cider drinker, would be able to tell you about the tannins and the flavours and the properties of the varieties used in the ciders. All I can say is that they induced a great sense of well-being, yet I woke up next morning clear-headed, refreshed even, and soon after sunrise I climbed through the orchards and meadows behind the farm, wading knee-deep through the dew-sodden grass, accompanied by a fluty chorus of birdsong. From the scrub at the top of the hill there were fine views over orchards, caravan parks, cider barns, fields full of sheep and, in the distance, the town of Paignton, which is where the Hunts' cider story began.

The cider business is now run by Roger's children, Richard and Annette. Richard had taken an unusual career path, studying English literature at Cardiff University before becoming a profes- sional rugby player – he has the towering height of a second row forward, if not enough weight for the modern game – with Bristol, Plymouth and Pontypool. In 2005, he decided to return home to make a go of the family's cider business. Annette, a pretty young woman a couple of feet shorter than her brother, had also been away from the farm, but always in the world of food and farm- ing, working as a jillaroo – a livestock hand – in Australia, then running a £40 million-a-year sandwich account for Samworth Brothers in Melton Mowbray and spending time on secondment to Tesco. Her marketing skills, allied to her brother's ambitious expansion plans, have transformed the Hunts' cider business.

'We were farming in Paignton in the 1770s', explained Richard, 'and a company had been set up to make cider by 1805.' The evidence is provided by some old bottles inscribed 'NP Hunt and Son'. Nicholas Prout Hunt was a yeoman farmer and his name appeared as a director on the Paignton Harbour Act, the first piece of legislation signed by Queen Victoria. Four generations of NP Hunt continued to produce cider into the early twentieth century. The last was childless and the company was taken over by a brewery, then sold on to Whiteway's Cyder. In 1915, Richard's great-grandfather became a tenant of Higher Yalberton Farm, a couple of miles down the road from where Richard and his family live now, and the current cider-making business is a direct descendant of the one he established. Roger, incidentally, had kept the cider business going, but for him it had never been a passion. He had made cider largely to keep his mother happy, as she had taken pleasure, for some sixty years, in selling it to the passing public from her kitchen window.

Wander through the orchards with Roger and he will point out trees planted by his grandfather, his father, himself and more recently, Richard. Old-fashioned orchards such as these, with trees of different ages often widely spaced, will give a yield of around four tonnes of apples an acre, a fifth of a modern commercial orchard designed to maximise yields. Most of the cider produced by the Hunts is made from either bittersweet or bitter-sharp cider apples. They include well-known commercial varieties such as Dabinett as well as older ones like Paignton Marigold, which the

Hunts believe they may have developed themselves. 'Raw cider apples taste disgusting, and you'd never think of eating them,' said Roger, 'but they make the best cider.'

The actual process of making cider at the Hunts' farm is much as it always was, with the apples being hand-picked by a couple of men – long-bearded, apparently – who come every year from Cornwall, then washed and crushed before the juice is stored in vats, where it ferments for a month or so. It is then racked off and left to mature, during which period the tannins soften and the flavours develop. By the spring, the cider is ready to be bottled.

When Richard returned to the farm some ten years ago, his father was selling just 1,000 gallons of cider a year. Over the next seven or eight years, as he developed his cider-making skills, he steadily increased production, which really took off once Annette brought her marketing nous back to the farm. During the season prior to my visit they had made 32,000 gallons of cider. Besides selling to pubs and restaurants, they also do good business at festivals and food fairs and they have developed close relationships with Wetherspoon and other retailers.

During recent years Richard has planted some five acres of new cider apple orchards; the Hunts have also built a large new cider barn – as lovely, in aesthetic terms, as any modern metal shed can be – and kitted it out with the latest machinery. This means they can now take full advantage of the expanding cider market. There is one company in particular to thank for the cider renaissance – the Irish cider makers, Magners. 'About ten years ago, they spent millions on a huge advertising campaign to make cider fashionable,' said Annette. 'They kick-started the idea of serving cider on ice as a summer drink. They made it look trendy and that attracted a new generation of cider drinkers.' Not only did the campaign help Magners increase sales, it helped every other cider maker, including the Hunts, who are now producing fruit ciders flavoured with strawberry and blackcurrant, as well as their more traditional fare, such as Percy's Pride and Mary Maud's Medium.

The night before we left, we went for a meal at a pub in Stoke Gabriel, a pretty village on the lower reaches of the River Dart, with Richard's parents, Roger and Christine. There was much chatter about Annette's forthcoming wedding, but we also talked about the changes which had taken place during Roger's lifetime. It was a familiar story. Besides making cider, his father used to grow cauliflowers and potatoes, and he raised turkeys, chickens, pigs and beef cattle. He used to take a truck to Paignton every week and sell produce direct to the townsfolk. 'Then the big supermarkets arrived in the 1970s, and they made it difficult for people like my father to compete,' explained Roger.

Nevertheless, the Hunts have adapted and survived. 'We always do our best to avoid dealing with the supermarkets,' said Roger. 'We like to go eyeball to eyeball with the customer.' For many years, Roger had dairy cows, originally South Devons and later British Friesians, but the cows have gone, to be replaced by caravans and motorhomes. At one time, subsidies encouraged Roger to farm quite intensively; now, subsidies encourage the exact opposite and the farm receives almost as much income from a Higher Level

Stewardship scheme – aimed, in particular, at retaining the scrub where rare cirl buntings breed – as it does from the Basic Payment Scheme. They no longer use fertilisers and a great profusion of wildflowers have returned to the meadows on the hillside. 'Our farm – it's almost like a nature reserve now,' said Roger with some satisfaction, and perhaps surprise.

The boom in cider sales has encouraged many farmers to establish new orchards. It has also attracted the attention of large brewers who are concerned about the flagging sales of beer. For example, in 2008 the Dutch company Heineken bought Bulmers, which has almost doubled the area under apple orchards in the last twenty years to keep pace with rising demand. However, it isn't always a question of big fish eating little fish. Thatchers is one of the well-known names to retain its independence, and it now accounts for around 6 per cent of the cider market in the UK.

'What's the secret of your success?' I asked Martin Thatcher as soon as we sat down in his smart and well-appointed office at Myrtle Farm, in the village of Sandford.

'A combination of science and hard work,' he replied, before adding that the company had been in the right place at the right time.

For many years, Taunton Cider had been the most successful producer in the West Country, its Blackthorn Cider dominating the local pub and restaurant market. Then the company underwent a series of acquisitions. 'I think they lost their West Country

Martin Thatcher in one of his orchards.

roots,' reflected Martin, 'and they began making a sweeter cider in the hope that this would have wider national appeal. But the West Country drinks dry cider and we were able to move into the market.' Within a short space of time Thatchers Gold had become the drink of choice in the region and it is now sold in supermarkets and pubs the length of the country. It's also the cider you drink if you go to sports matches in the West Country; Thatchers sponsors Bath Rugby Club, Somerset Cricket and Yeovil Town football.

The company's growth has been astonishing and although it seems to have retained the ethos of a family business, rooted in North Somerset, it has all the accoutrements of a successful industrial concern, including a charming PR consultant, Penny Adair, who prompted Martin every now and again if modesty got the better of him. 'Martin and his dad John have been incredibly innovative,' she said at one point. 'They were the ones who pioneered the idea of making single varietal ciders.'

The company recently produced what Martin called 'our periodic table', a beautifully designed chart of twenty-four varieties of apple, grown on some 440 acres of orchards around the village of Sandford. 'The first single varietal cider we made used Katy,' he explained, pointing to a red apple on the left-hand side of the chart, at the eating-apple end of the spectrum. 'Then we followed that by making other single-varietal ciders with Tremlett's Bitter, Dabinett and Somerset Red Streak.' These three were at the cider-apple end of the chart, with much higher levels of tannins. 'Producing the single varietal ciders gave us an opportunity to show how varied ciders could be, and they helped to distinguish us from our competitors.'

I asked whether cider drinkers were really that discriminating.

'People have become more discriminating. They are now much more interested in what they're eating and drinking than they were in the past, and how it's produced and where it comes from.' During recent years, Thatchers has been experimenting with a range of different varieties on different soils, using some of the 458 varieties acquired from Long Ashton Research Station, which before its closure was the only centre of science for cider in the country. 'We haven't got to the stage of talking about a *terroir* for cider yet, but we are moving in that direction.'

There isn't a direct English translation for *terroir*, a word used by French wine makers to characterise topography, soils and climate, and the taste and flavour they give to the wine. But listening to Martin talking about different varieties reminded me of the sort of conversations I have had with French *viticulteurs*. He described Dabinett, his favourite variety, as soft, round and mellow; Vilberie has so much tannin it is almost chewable; Somerset Redstreak is peppery; Tremlett's Bitter is robust and challenging.

Before we headed out to look at one of the orchards, Penny suggested to Martin that he should tell me about his experience as a Nuffield scholar. Every year, the Nuffield Farming Scholarships Trust provides some twenty individuals with the means to travel abroad to broaden their horizons and help them develop new ideas which will benefit the farming industry. In 2005, Martin was awarded a scholarship to look into the health benefits of non-citrus fruit juices and he travelled to Australia, the United States, China and Poland.

'It was the best bit of education I'd ever done,' he said bluntly. 'It helped me to understand why some companies are successful and do better than others.' He cited, as an example, Ocean Spray and their cranberry juice. 'What I realised when I looked at their operations is that everything they did was just that little bit better than their competitors. They also worked out the best way of marketing the juice, focusing on its health benefits.' It is claimed that drinking cranberry juice can help tackle urinary infections. Tapping into this market has done wonders for Ocean Spray's sales.

'I came back and started thinking how could we do things differently,' said Martin. Besides moving into the production of single-varietal ciders, he decided to abandon one side of the business, pressing other people's apples and selling the juice to other companies. Thatchers also invested heavily in advertising on TV, playing on the notion of family, community, rusticity and golden sunsets: 'One taste of Thatchers Gold and you'll know it was made by people who care.' Martin approvingly quoted a colleague who said that having a great product and not advertising would be like winking across a dark room at someone you fancy.

As we wandered around one of the orchards, we talked about branding. Martin told me about a farm he had visited several years ago that grew coriander. They were very successful for four or five years, but they failed to create an identifiable brand specific to their farm, and now other farmers in the neighbourhood were copying them and taking a slice of the market. 'Farmers need to think about the consumers and create something special that consumers want to buy,' suggested Martin. 'They need to take advantage of nicheness, if there is such a word, and create something which is hard for their competitors to copy. For some parts of the agricultural industry, such as dairy farming, that's very difficult to do. But we can do it with cider.'

It was now October. The autumn colours were creeping blotchily across the landscape and the apple harvest was well underway. There were robins and chaffinches singing in the orchard and we stopped to watch two roe deer grazing among the apple trees. This orchard, like the others belonging to the Thatchers, is run in a very scientific manner, with straight rows of even-aged stands, rather than trees of different ages planted further apart, which was the old-fashioned way. At the time of my visit, the company employed 183 staff, just six of whom worked in the orchards. 'We have introduced a massive amount of automation, both in the orchards – everything is harvested mechanically – and in the processing side of the business,' explained Martin. Besides using apples from their own farms, Thatchers also takes apples from some forty farmers in Herefordshire, Gloucestershire, Worcestershire, Somerset and Devon. From the farmer's point of view this makes good economic sense, as apple orchards can be more profitable than arable crops, especially when companies like Thatchers provide a guaranteed market.

One of the things that impressed me during my travels in the West Country was the environmental friendliness of cider production, whether it was a traditional system, such as the one run by the Hunts in Devon, or a modern business like Thatchers, where the apples are harvested by machines costing several hundred thousand pounds apiece. The company uses as little fertiliser and pesticide as possible, adopting an approach known as integrated pest management: getting nature to do the job of control, rather than chemical sprays. The orchards are full of wildlife, and indeed they depend on wildlife, whether it is tits and robins to control insect pests or bees to pollinate the blossom.

Should this influence my drinking habits? It probably should. Whenever I can, I buy beef that comes from native breeds that have been reared on grass, rather than from animals which have been fattened indoors on cereals. My choice is influenced to a certain extent by taste, but I favour grass-fed beef because of its environmental benefits, more about which in Chapter 7 on the

Scottish Borders. If I were to apply the same logic to my drinking, then I should favour cider over beer. Drink cider and you are helping to provide a future for our cider apple orchards, which possess many of the benefits – in terms of supporting wildlife, sequestering carbon and maintaining healthy soils – associated with deciduous woodland. Drink beer and you are supporting the production of barley, most of which is grown as a monoculture, and often dosed with agricultural chemicals.

Study the long-term figures for fruit production in the UK and you might think, at first sight, that this is a depressing story. Measured by volume, we only produce 10 per cent of the fruit we eat. In some ways, that's not surprising, as we import huge quantities of bananas and oranges, which together with dessert apples account for 80 per cent of the fruit consumed. In terms of indigenous fruit – both top fruit such as apples, pears, plums and cherries, and soft fruit such as strawberries, raspberries and blackberries – the figures are better: our farmers and growers supply 40 per cent of our needs. However, there is the potential to significantly increase production, and the soft fruit sector is beginning to show the way. According to British Summer Fruits, an industry body, demand for strawberries has been growing by 12 per cent a year and other soft fruit by 25 per cent or more a year.

When I was growing up in the 1960s, we ate vegetables in season, which meant greens in summer and root crops in winter, and the same applied to fruit, to raspberries and strawberries, gooseberries and blackcurrants, blackberries and rhubarb. In those days, and right up till the early 1990s, British strawberries were only available for about six weeks each summer and if it was a poor spring or summer scarcity led to high prices. Now, thanks to the use of polytunnels – elongated greenhouses with transparent polythene stretched across a metal frame – the season lasts for about twenty weeks, and strawberry production has risen by over a third in the past five years. Some of the new varieties are almost 2 inches long by an inch wide, but they are not necessarily the tastiest.

'These are not just any strawberries, they are Little Scarlets – the king of strawberries!' said Chris Newenham, the joint

Chris Newenham in a strawberry polytunnel at Tiptree.

managing director of Wilkin & Sons Ltd in Tiptree, Essex, the largest vertically integrated fruit and jam making business in the country. We were squatting on our haunches in a field next to the Blackwater Estuary, eating berries that had ripened since the pickers had done their work. It was one of those rare days, almost too perfect even for a fine English summer, with the sun beating down from a cloudless blue sky. A softly spoken Irishman, brought up on a farm in County Cork, Chris clearly had a passion for his job, and for Little Scarlets in particular. 'No other strawberry tastes this good,' he said.

All modern strawberry cultivars are descended from a cross between *Frigaria virginiana*, the Little Scarlet, a native of northeast America, and *Frigaria chilonsis* from Chile. The hybrid was developed in Brittany and in market gardens it swiftly replaced the woodland strawberry, *Frigaria vesca*, which was first cultivated in Britain in the early seventeenth century. This, presumably, was the variety which the Bishop of Ely was referring to in Henry V: 'The strawberry grows underneath the nettle/ And wholesome berries thrive and ripen best/ Neighbour'd by fruit of baser quality.' In appearance, it was much like the Little Scarlet, about one-fifth the size of a modern strawberry.

The first Little Scarlet plants were brought to England from America in the early 1900s, and among the first to grow them here were the Wilkin family, who had been farming in and around the village of Tiptree for some 200 years. Tiptree – this is how the company refers to itself – is thought to be the only commercial business to grow the strawberry, and its Little Scarlet Strawberry

Conserve has gained a worldwide reputation as one of the finest of its kind.

Relatively poor soils and mostly meagre rainfall have meant that the company has always been searching for ways of improving its strawberry production, most recently by taking soils and the weather out of the equation. As we headed away from the estuary and back towards the main farm and the jam factory, Chris led me though the evolution of Tiptree's cultivation methods. A short distance from the field where we tasted the Little Scarlets was a polytunnel where the strawberries were grown in bags filled with coir, coconut husk fibre imported from Sri Lanka. A trickle irrigation scheme keeps the coir damp and delivers nutrients. A similar arrangement had been established in other polytunnels on long thin metal tables, waist-high for ease of picking. Eventually, we arrived at the latest development: a huge 15-foot-high polytunnel with a two-tier arrangement of irrigated tables worked on a pulley system.

'Under this form of management, there are up to 200,000 plants per hectare, compared to some 50,000 when grown in open fields,' explained Chris. There are many other advantages too. With enclosed polytunnels, the plants are protected from heavy rains and cold spells and they are less likely to be attacked by pests and diseases than strawberries grown in open fields. The drip irrigation systems in the polytunnels are also very efficient. Indeed, this emphasis on environmental efficiency was one of the reasons why I wanted to visit Tiptree.

In June, I had spent two days at Cereals 2016, the largest annual technical event for arable farming, held at a farm in Cambridgeshire. All the latest combine harvesters, seed drills, sprays, ploughs and other cropping paraphernalia were on gleaming display. But there was a less showy, and equally important side to the event: Cereals 2016 attracted a range of organisations which were promoting technologies, or ideas, designed to make the growing of crops more efficient, sustainable and environmentally sensitive. One of these was Linking Environment and Farming (LEAF), an organisation – or movement – established in 1991.

I chatted briefly with Caroline Drummond, LEAF's chief executive, and she explained how the organisation is dedicated to helping farmers and growers improve their soil management and fertility, reduce pollution, practise good livestock husbandry and focus on energy efficiency: in other words, do everything they can to farm in a sensitive and sustainable manner. Around a third of all the fruit and vegetables produced in this country is now LEAF-accredited.

'We are enthusiastic supporters of LEAF,' said Chris. 'It is a common sense philosophy, and it helps us think about ways to improve what we are doing. In the recent past we've had some fruit under organic production, but that's more fundamentalist than LEAF.' By which I think he meant that it was more prescriptive, less flexible. By adopting LEAF's integrated farm management approach, Tiptree has significantly reduced its water consumption. Winter rain is harvested in small reservoirs and summer rain is captured in gutters and recycled. The company extracts no groundwater. 'In the past, we used to spray prophylactically to control pests and diseases,' continued Chris. 'Now, we only spray chemicals as a last resort.' Whenever possible, the farm uses biological methods of control, for example by introducing a predatory mite, *Phytoseiulus persimilis*, to control red spider mites on strawberries. This sort of benign pest control is becoming increasingly popular.

Under Chris's management, Tiptree has benefited from a variety of stewardship agreements which have helped to fund the planting of over 5,000 native trees – none of which are producing commercial fruit – and some three miles of hedgerows. The fields near the estuary are popular with wading birds in winter and brown hares throughout the year. On one occasion, hares wiped out six hectares of strawberries. 'That was an expensive year,' mused Chris.

Great attention had also been paid to reducing waste and improving efficiency in the factory, which looks like the sort of place that would have delighted Heath Robinson, with its vats full of bubbling fruit and elaborate machinery for bottling jams, chutneys and pickles. Environment Manager Kevin Townsend took me on a brisk tour. Today, he explained, was black cherry day. However, we also passed groups of women peeling Seville oranges and preparing scotch bonnet peppers for chilli chutney. Just ten years ago, the company recycled 70 per cent of its waste on site; last year, it recycled 91 per cent. Fruit waste is transformed into energy at an anaerobic digester nearby, and nothing goes to landfill now. The company has also reduced its consumption of water and energy. 'We are using 25 per cent less water than we used just two years ago, and last year we reduced our gas consumption by 13.6 per cent,' said Kevin proudly.

I am sure you can hear a similar story of prudence and efficiency in many other manufacturing sites in this country. But I doubt whether you'd hear as much praise for the management. 'This is a fantastic company to work for,' said Kevin as we passed bottling machines which can churn out 300,000 little pots of jam a day. He told me a story about a man called David who had worked here for fifty years, retired, spent two weeks at home getting bored, then returned to work full time. Some twenty-five years ago, the company set up an Employment Benefit Trust, which provides an annual allocation of shares to all employees. When they retire, they cash in their shares, which are then passed on to new employees. The company also has about seventy houses in the village, providing subsidised rent to the men and women who work here.

There is a black-and-white photograph outside Chris's office of Mrs Barnard, aged ninety-four, picking raspberries at Bound's Farm in 1920. In those days, it was the indigenous population who did the picking, and indeed so strong was the tradition that schools in the area used to have a long break during the summer term so that teachers and children could make some money in the fields. In the late 1950s and 60s, it was the Irish and Scottish who came to pick fruit, followed in later years by Germans, French and Dutch, then Spanish and Portuguese, Ukrainians and Mongolians, Africans for a while – often students studying in Turkey – and finally Eastern Europeans.

Nevertheless, Tiptree continues to welcome English pickers too, and still has a field with basic facilities where they can park their caravans or pitch their tents. 'If you had come twenty years ago, there would have been up to 300 caravanners here,' said Chris as we drove past the site. 'Now, we get thirty or forty, mostly the older generation who enjoy the work, even though it's hard work picking.' Unfortunately, the national minimum wage has made life worse, rather than better, for some of these old pickers, who are paid – or were paid – piecework. The company is now legally obliged to pay them the minimum daily wage, which may exceed the amount they can earn through piecework. 'We simply can't keep making up their earnings to the minimum wage, so now it's three strikes or they're out,' says Chris. 'It's a real shame as many of them come for the social life, to chat in the fields while they pick fruit, and they're quite happy just to earn £30 a day.' One of the pickers, a schoolteacher from Doncaster, had been coming for forty-eight years.

In 1895, the company produced some 200 tonnes of fruit a year and employed 400 pickers on a seasonal basis. Nowadays, Tiptree grows 1,250 tonnes of strawberries, cherries, plums, raspberries, mulberries, quince, meddlars and other fruits for the fresh market and making jams. It has a full-time staff of 375, 85 per cent of whom are British, and 350 or so seasonal staff, virtually all of whom are Eastern Europeans.

'I simply cannot speak highly enough about our Eastern Europeans,' said Chris when he took me to see the company's international farm camp, which has fifty-five fixed caravans, each sleeping six, and a large community hall equipped with a laundry, pool and table tennis tables and a gym. A couple of years ago, the Croatian ambassador to the Court of St James signed his name on one of the walls. As a young man, he had come here as a fruit picker.

In 2002, Chris was working on a fruit farm in Hampshire. One day he went to Brockenhurst station to pick up a couple of Bulgarians who had come for seasonal work. 'They were wearing black suits and black shoes, and I thought to myself: they won't last long,' recalled Chris. 'I couldn't have been more wrong. They were the best pickers we ever had.' One of them, Andrey Ivanov, is now the farm manager at Tiptree. His brother Veso is in charge of recruitment.

We found Veso in the office at one end of the community hall and Chris left me to chat to him for a while. In the past, Veso explained, around 95 per cent of the Eastern Europeans who came here for the fruit-picking season would return the following year. But he was worried that the recent Brexit vote would change that. The pound had already gone down, and this effectively meant that the Romanians and Bulgarians were getting less pay when converted into their own currency. 'Some of them will look for work in other countries like Spain or Germany,' said Veso. He said many of the migrant workers were worried about whether they would be allowed into the country to work in future, either with or without visas.

Over lunch in the Tiptree Tea Room, I asked Chris about his views on Brexit. He was clearly distressed by its implications. 'It's as though the lid's been taken off a pressure cooker,' he said. Since the referendum vote, just three weeks before my visit, some of the Eastern Europeans had been told to go home by locals on the street. He thought Brexit would make it harder to recruit labour. The fall in the value of the pound also meant that certain products, such as imported glass jars, were now more expensive, although it was good for exports. 'We will play the hand we've been dealt, but it's going to be challenging.' I asked whether he would be able to find local labour if it became harder to hire Eastern Europeans. He shook his head. In his experience, most young English people were not prepared to work in the fields. He had tried recruiting locally in recent years, but with little success.

Several months after I visited Tiptree I spent time in the Caribbean on an assignment which involved, among other things, talking to chefs about their efforts to source local produce, rather than rely on foreign imports. Kirk Kirton, the executive chef at the Fairmont Royal Pavilion in Barbados – the sort of place you'd love to stay if only you could afford it – reeled off all the products he was now buying from local farmers: chicken, sweet potatoes, asparagus, pumpkins, tomatoes, lettuce, fresh herbs, tropical fruits. But there were some essential items that would always have to come from abroad. One was Tiptree's orange marmalade; another was its Little Scarlet Strawberry Conserve. 'If you're going to serve afternoon tea in a place like this,' he said as we gazed out over the placid blue sea, the temperature nudging 30°C, 'then you have to have strawberry jam and clotted cream, and it's got to be Tiptree's Little Scarlet.'

5 The Art of Growing Vegetables

I ended the last chapter in the Caribbean and I shall begin this one in the Pacific. If you have had the good fortune to visit either of these regions, one of the first things you will have noticed is the epidemic of obesity. During recent decades, Pacific islands, like most in the Caribbean, have seen a dramatic increase in Type II diabetes, heart disease and strokes. Around 70 per cent of all deaths on Samoa and on many other Pacific islands are now caused by non-communicable – mostly diet-related – diseases.

Talk to any Samoan over the age of sixty and they will tell you how the islanders once thrived on a diet of fresh fish, root crops, vegetables and fruit, all locally harvested. Obesity was almost unknown when they were young. Now, most of the food is imported – not least because it tends to be cheaper than home-grown food – and the younger generations have developed a taste for New Zealand mutton flaps, American chickens the size of small turkeys, deep-fried turkey tails and processed goods made out of wheat flour. Consumption of vegetables and fruits is now pitifully low, especially among the poor.

The situation in this country is not quite so dire, but you only have to look around you – for some reason motorway service stations provide a good visual window on our ill health – to see that we have a crisis of our own: 63 per cent of adults were classified as overweight or obese in 2015 and we rank second in the obesity league table in Europe. The UK now spends £14 billion a year on treating diabetes and complications related to the disease, which is roughly the amount allocated for foreign aid. The financial costs related to a poor diet tell just part of the story. You can't put a figure on the physical and emotional misery caused by eating too much of the wrong things and not enough of the right. The solution is obvious: we need to reduce our consumption of unhealthy foods and increase the amount of fresh fruit and vegetables in our diet. According to the government's Eatwell Guide, at least one-third of our daily diet should comprise fruit and vegetables. At present, they comprise less than a quarter. Just one in five people consume the recommended five or more portions of fruit and vegetables a day.

While our levels of cellulite and cholesterol head in one direction, the area of land devoted to growing vegetables has gone in the other, falling by 26 per cent between 1985 and 2014. Nevertheless, the sector still generates 20 per cent of farm-gate value – the price of the product when it leaves the farm, after marketing costs have been deducted – from just 3.5 per cent of the land suitable for growing crops. In 2015, UK farmers produced 2.5 million tonnes of field vegetables and a further 310,000 tonnes of vegetables under glass or polythene, satisfying 57 per cent of our needs.

Potato crisps and chips are not considered part of a healthy diet, but that's largely because of the way they are manufactured and cooked. However, potatoes have much to recommend them, especially when eaten boiled or baked, as they are a good source of vitamin B6, vitamin C, potassium, phosphorus and fibre, among other things. I have included potatoes in this chapter for two reasons. For one thing, they are botanically classified as a vegetable, although they are considered as an arable crop for statistical purposes. For another, the cultivation of potatoes, like the cultivation of vegetable crops, requires a high degree of skill. Potato growing also accounts for almost as large an area of land as all other vegetables put together and double the volume.

My experience of working with potatoes and green vegetables has been limited and tedious. In the early 1970s, I picked potatoes by hand with an extended family of gypsies who lived in caravans on the A1 near Scotch Corner. It was tough, back-breaking work, and I was a slow and inept picker. A few months later, during a ferociously cold December, I spent a weekend picking Brussels sprouts – it was like twisting frozen billiard balls off iron bars – and that was hard, miserable work as well. More enjoyable were the warm days spent hoeing turnips in June. Hoeing by hand can be a monotonous task, like pulling wild oats in wheat fields or clearing ploughed fields of stones in winter, but in those days you always worked in a gang, so you were never lonely.

One of the people I picked vegetables with – I can remember him in the potato fields, if not with the Brussels sprouts – was Harry Marwood, who was the head tractor driver at Clifton Castle, in Wensleydale, when I was a farm student. We have remained good friends ever since and he offered to take me to see potato growers in Yorkshire before I headed further south, first to visit an old-fashioned vegetable enterprise near Goole which supplies the grocery trade, and later to spend time with two major vegetable growers who supply our supermarkets.

For the past twenty-odd years, Harry, who is now an absurdly fit-looking seventy-year-old with a full head of hair, has provided advice to farmers who grow potatoes for McCain Foods, the largest processor of frozen chips and French fries in the country. Prior to that he had some thirty years of experience on arable farms, growing cereals and various other crops in Yorkshire and Lincolnshire. As we headed from his home near Wetherby towards York, we chatted about some of my experiences over the past few months. I mentioned that many of the farmers I had met,

Harry Marwood introduced me to the world of potato farming.

especially those involved in the dairy industry, had claimed that growing cereals like wheat and barley didn't require any great skills, and that cereal farmers only worked in fits and starts, with much time on their hands in winter.

'Well, they've got a point,' he replied. 'You could grow cereals with your eyes closed and almost no labour. It's simple compared to growing potatoes or vegetables.' Growing potatoes involved a high degree of risk and the sort of capital cost that didn't bear thinking about, as I was about to find out.

Tony and Linda Beckett's farm in the village of Long Marston lies at the heart of some of the best land in the country for growing potatoes, on a belt of limestone running down the western side of the Vale of York. We talked, at first, about the great changes that had happened over the last fifty years or so. When Tony left school, in 1968, there were ten farms in the village which grew potatoes; now there are just two. In 1968, the Becketts grew 10 acres of potatoes, which were harvested by hand – the Irish were the best pickers, according to Tony – and stored loose in a barn, where they were protected from winter frosts by straw bales. Today, the Becketts produce some 9,000 tons of potatoes on 500 acres and make use of highly sophisticated planting and harvesting machines.

When Tony began work on the farm, they could plant about 4 acres in a day; a modern potato drill can plant 25 acres. Instead of selling potatoes in bags, as they used to, the harvested crop is now lodged in purpose-built stores with elaborate ventilation and heat control systems. When Tony was a young man, most of the potato crop in the UK was sold fresh. 'Do you remember when we were young, a bag of crisps used to be a big treat?' he asked. Now well over half of all the potatoes grown in the country are processed into crisps and chips.

Instead of selling to local wholesalers, as they did in the past, most large-scale potato growers work under contract to companies like McCain and Walkers. 'There's too much risk involved in growing potatoes for most growers *not* to have a large acreage contracted,' said Harry, who for many years provided technical advice to the Becketts on behalf of McCain. 'You can sit down in winter, when the snow is on the ground, and you haven't a clue what the weather is going to be like in the summer, and what sort of crop you'll get. If you have a contract, then you can fix a price in advance and eliminate some of the risk.' McCain works with the growers to choose the varieties to suit the prevailing soil types and growing conditions and provides advice on fertilisers and agronomy.

'My father always used to say that if you bought a new combine, it would be out of the profit of potatoes, not grain,' said Tony. 'You can make much more money from potatoes than cereals, but you can lose more as well. There is a lot of risk in this business.' If you have a flood, you can lose the entire crop, and if it is exceptionally wet in autumn, you may not get into the fields to harvest the crop. If you have a contract with McCain to supply, say, 1,000 tonnes and you only harvest 700 tonnes, you will be 300 tonnes short. If you had been expecting £140 a tonne, you would be £42,000 worse off than you anticipated when you signed the contract. But at least you have the surety of McCain taking the rest of your crop.

Harry told me about one McCain grower in Yorkshire who spent £500,000 on a large refrigerated potato store during the summer of 2012. Unfortunately, a very wet autumn followed and a large acreage of potatoes was left in the ground to rot. The grower fell disastrously short of his contracted tonnage. The bank called in the money he had borrowed to build the store and he had to sell land and machinery. He was unlucky: if the wet autumn had come three or four years later, he would already have paid off a good proportion of his debt to reach a position of relative financial safety. 'That's why this is such a risky business – the weather can make or break you,' said Harry.

The second farmer who Harry took me to see operated on a large scale in and around the village of Kexby, near York. As we made our way to the local pub for lunch, Andrew Etty, a forthright and articulate man in his late forties, explained that over the past few decades the farms here had become larger and livestock had largely disappeared from the fields. Some farmers were now doing what's known as bed-and-breakfast pigs, providing indoor accommodation for somebody else's pigs and feeding them till they are ready for slaughter. The owners, rather than the farmers, are taking the risk. 'But I'm more of a gambler,' added Andrew with relish.

Every day he studies the weather forecast, frequently looking as far afield as the American Midwest, to see how climatic conditions might affect the yields of maize and other crops, and therefore future prices for crops like feed wheat. While many farmers sell their crops soon after they have harvested them, Andrew keeps his in storage and only sells when he considers the time right. He also follows fluctuations in the currency market, which can lead to the price of wheat or oilseed rape varying by £10 to £15 a tonne in a single day. To survive in future, he suggested, farmers will need to act like city traders as well as crop growers.

If the weather is a major preoccupation for potato growers, so is disease. The Irish famine of the 1840s, which led to the death of over a million people and the emigration of another million, was caused by the failure of successive potato crops following infestations of late blight. This is still the most threatening of all diseases for potato growers and the vast majority rely on fungicides to keep it at bay. 'Thirty years ago we would spray for blight five or six times a year,' said Andrew. 'Now, we have to spray every five or six days. Mother Nature adapts and we are in a perpetual arms race with diseases like blight.'

It was the same story with Tony Beckett. 'In the past, we used to say that we should start spraying for blight on Yorkshire Show day, in mid-July,' he said. 'But now we will have sprayed six times by then. I find it amazing that we have to spray as often as sixteen times in a season. The strains of blight seem to be getting more and more vigorous.'

One solution, he suggested, could come from the introduction of genetically modified (GM) blight-resistant potatoes. Researchers have spent years trying to develop resistant varieties by crossing commercial varieties with wild – inedible – potatoes from South America that are naturally resistant to the disease. Doing this using conventional plant breeding techniques has proved difficult. However, scientists in Britain have successfully added a gene from a blight-resistant South American potato to Desiree potatoes, making them fully resistant to late blight. Norwegian scientists have also used genetic modification to create blight-resistant potatoes. The first varieties of GM potato are likely to be planted in the United States in 2017.

Current EU regulations mean that there is little prospect of genetically modified potatoes being planted commercially in this country, or elsewhere in Europe, in the near future. A vocal and well-organised coalition of environmental organisations has done everything it can to convince the authorities, and the public, that GM crops are a threat to both the environment and human health. However, a recent report published by the National Academies of Sciences, established in the United States by Abraham Lincoln to provide independent scientific guidance, suggests that GM crops have been unfairly maligned. The report found that researchers had identified 'no differences that would implicate a higher risk to human health' from eating GM foodstuffs than eating non-GM foodstuffs. Researchers had also found no evidence that GM crops caused any more environmental problems than non-GM crops. It remains to be seen whether evidence such as this will convince the doubters in this country. In the meantime, the vast majority of potato growers will carry on tackling blight by drenching their potatoes with fungicide.

So, you might ask, why not grow organic potatoes without using pesticides? After I got home, I rang Harry to ask whether he had any experience of organic potatoes. He said he had worked with one farmer in South Yorkshire who had tried to produce potatoes organically, but with little success. Of course, it is possible to grow organic potatoes, but yields are much lower than they are for conventional potatoes. 'In any case,' said Harry, 'if this country had to rely entirely on organic potatoes, they would either be too expensive for consumers to buy or too unprofitable for farmers to grow, or both.'

The most fertile agricultural land in Britain is concentrated in a long strip, some 50 miles from east to west at its widest, stretching from the Vale of York to South Essex. Of this, the very best land is found around the Wash and in the Fens and along the banks of the Humber Estuary and the River Ouse. If you want to grow vegetables, this is one of the best places to come, and this is where I headed after I left Harry and his potato growers.

I arrived at Greenside Farm one breezy July morning. The farmhouse itself was a handsome building overlooking the green in the village of Rawcliffe, which sits on the Ouse floodplain to the east of Goole. Beryl Sykes invited me in for coffee and before we went to look at the crops we chatted about the history of the farm. From time to time we were joined by her son Patrick, an energetic young man who gave the impression

Beryl Sykes and her husband Alan.

that he was always on the move, but it wasn't until midday that her husband Alan appeared. He had been out in the fields, hoeing all morning, and he had returned for what Beryl described as the highlight of their day, which was lunch, or dinner as it is called in the north.

It would have been rude to ask their ages, but I imagine Alan, a fit, craggy figure with grey hair, bushy eyebrows and a weathered complexion, was in his seventies, and Beryl quite a few years younger. They had first met when her father had taken on the tenancy of a 40-acre county council farm in Rawcliffe. She was then eight years old. Alan lived with his family on the farm next door, where we were now, and they married in 1967.

In those days, the county council had around thirty tenant farms in one large block, carved out of an estate which it had bought after the Great War. 'There's only about fourteen or fifteen of us tenant farmers left now,' explained Beryl. 'Whenever one of the farms becomes free, the council will sell off the house and amalgamate the land with another holding.' They were hoping that Patrick would be able to take over the tenancy of this farm, but nothing had been agreed yet. East Yorkshire County Council still has a significant number of tenant farms, but many councils are divesting themselves of their agricultural properties as a cost-cutting exercise. Families like the Sykes are among the dwindling number of people who have benefited from a system of land tenure which has helped thousands of young men and women get on the farming ladder.

We had just finished Beryl's home-made salmon plait, a delicious dish accompanied by salad and vegetables from the farm, when Alan said: 'So you're doing a Cobbett, are you?'

I wasn't used to farmers mentioning William Cobbett, the early nineteenth-century pamphleteer, politician and author of *Rural Rides*. 'So you've read Cobbett, then?' I replied.

'My mother was a Cobbett on her dad's side, going back four or maybe it was five generations to one of Cobbett's brothers. I could see Cobbett in my grandad's face.'

Alan wanted to know whether I had read Cobbett's advice on how to choose a wife. I said I hadn't and he went in search of a copy of *Advice to Young Men, and (incidentally) to Young Women*. He indicated the passage he was referring to, which suggested that when looking for a wife a young man should remember the old adage, 'quick at meals, quick at work'. Never mind the fancy needlework, suggested Cobbett: 'Get to see her at work upon a mutton chop, a bit of bread and cheese; if she deals quickly with these you have a pretty good security for that activity, that steering industry, without which a wife is a burden instead of being a help.'

'And that's how I chose Beryl,' said Alan, grinning broadly.

In fact, this looked to me like a partnership of equals. 'We are a team,' said Alan. 'Everything I can't do, Beryl can. Same with Patrick. I know how to grow vegetables. Patrick knows how to cut them. It wouldn't work so well if any one of us wasn't here. It just wouldn't be the same.'

There had been considerable changes on the farm since he was young. For one thing, they always used to have livestock and only gave up cattle in 1988, when Alan's father died. They also had sheep for a while. 'We enjoyed the livestock and we miss the muck now,' said Beryl. As well as supplying them with fertiliser, the muck helped to keep the moisture in the vegetable fields. They now use green manure from crop waste and increase the organic matter in the soil – a good silty warp, as Beryl described it – by ploughing in chopped-up straw.

The Sykes have around 250 acres of land – some tenanted, some owned – and at any one time there will be up to 40 acres of vegetables, with the rest devoted to cereals. Beryl reeled off a list of the vegetables they were growing this year: leeks, cabbages, cauliflower, swedes, turnips, pumpkins, squash, courgettes, sweetcorn, dwarf beans, runner beans, vining peas. The peas are sold under contract to the nearby Birds Eye factory – they are machine-harvested by contractors at night and must be frozen within a few hours – and most of the rest goes to wholesalers in towns like Hull, Doncaster and Sheffield.

'We are niche producers,' said Patrick. 'A wholesaler may ring up and ask for ten boxes of beets, ten boxes of carrots and ten boxes of something else. That's the way we operate.'

They are part of a dwindling breed. In the mid-1980s, approximately half of the vegetables traded in this country were sold by farmers like the Sykes to some 10,000 wholesale merchants who supplied markets and greengrocers. Today, just 15 per cent of vegetables go into the grocery trade; the rest are bought by supermarkets and processors.

I asked how they arrange the price.

'When we are cutting – mostly in the early morning – we often don't know what we're going to get,' said Patrick. 'A lot of people laugh at us for that, but Beryl knows the wholesalers and she has built up a good relationship with them.' The Sykes tend to supply their best payers first. When they have plenty of produce they look for new customers and may sell at a cheaper price. In times of shortage, on the other hand, they can demand higher prices, and they set the price themselves when selling to small customers, such as farm shops and garden centres. They also have a stall outside the farmhouse where local people can buy vegetables in season.

They clearly love this way of life. It is hard, certainly: most mornings they get up at dawn and during the cereal harvest

Patrick can be working in the fields as late as 10 o'clock at night
– to the chagrin of some people in the village. They harvest vegetables by hand with a knife, doing most of the work themselves,
and the crops are weeded using a long-handled hoe. 'I have enjoyed
hoeing ever since I was fourteen or fifteen,' said Alan, before heading back to the fields after lunch to do more of the same.

Farmers always talk about the weather, but the Sykes, and the
other vegetable farmers I met, pay more attention to it than most.
'We watch it all the time,' said Beryl. It wasn't just a question of
needing dry weather or wet weather, hot weather or cold weather,
but the right weather at the right time. 'When you're packing in
cardboard boxes, you need it to be dry, but if it's wet, it's good for
getting carrots out of the ground. The wet weather last week will
have suited our leeks, but we could do with some hot, dry weather
now for the courgettes and pumpkins.'

Horticulture, like the pig and poultry industries, has never
received any production subsidies. Growers have always lived
by the laws of supply and demand and the price for vegetables
fluctuates accordingly. For example, during the month of July
2016, when I visited the Sykes, the wholesale price of asparagus
increased by 22 per cent, reflecting end-of-season shortages; the
price of broad beans fell by 12 per cent, reflecting an increase
in domestic supply; the price of red onions fell by 21 per cent
due to an increase in imports; the price of Kentish cauliflowers
rose by 49 per cent as supplies dwindled; and the price of runner beans dropped 38 per cent as supplies increased and demand
fell. This doesn't mean that growing vegetables is an inherently

risky business as far as the market is concerned: depressed prices for some will be compensated by increased prices for others. Bad weather, on the other hand, can have a serious impact on families like the Sykes who grow their vegetables outdoors rather than under glass.

After I left Yorkshire, I headed south to two great estates in East Anglia: the Elveden Estate near Thetford, owned by Lord Iveagh; and the Houghton Estate, once the home of the first British prime minister, Sir Robert Walpole, and now of the 7th Marquess of Cholmondeley, in North Norfolk. I had been pointed in the direction of these two estates by Caroline Drummond, who I mentioned in the previous chapter. You may recall that Tiptree, the Essex fruit business, is a keen member of Linking Environment and Farming (LEAF), which is run by Caroline; so are Elveden and Produce World, which has a large organic vegetable business on the Houghton Estate.

LEAF is very much a child of its times. When it was established some twenty-five years ago, there were serious concerns about many aspects of modern farming. For example, there was clear evidence that the use of pesticides to protect crops from fungi, bacteria, insects, weeds and rodents was having a serious impact on both the environment and human health. Besides killing the target species, pesticides frequently lead to the death of other species, including insects which prey on crop pests, and plants which provide food for farmland birds such as partridges, skylarks and corn buntings.

The indirect impact of pesticides on the environment can also be serious, for example when they are inadvertently sprayed on hedgerows or leached into watercourses. We now have a strong regulatory system and the worst offenders, such as DDT, have long since been banned. Indeed, since 2001 farmers in the EU have lost over half of the 850 active ingredients registered during this time, with the list of banned substances steadily increasing. As a general rule, the less pesticides that farmers use the better, both in terms of cost to them and impact on the environment. One of LEAF's aims has been to encourage its members to reduce their use of agrochemicals by adopting practices that come under the heading of integrated pest management.

There are now almost 1,000 businesses certified with the LEAF Marque, an environmental assurance scheme recognising the sustainability of farm produce. Over 280,000 hectares of crops are LEAF assured in the UK, with 42 per cent of the area being devoted to horticulture. By 2016, 85 per cent of all leeks, 76 per cent of beetroot, 70 per cent of lettuce and 59 per cent of asparagus grown in the UK qualified for the LEAF Marque. LEAF has also played a significant role in introducing the public to the farming industry. In June 2016, 382 farms participated in Open Farm Sunday, which attracted 261,000 visitors, over a fifth of whom had never been on a farm before.

I came to see Andrew Francis, senior farm manager at Elveden, the day after Theresa May had been appointed prime minister.

'You could put a farmer in Number 10,' he said as soon as we had shaken hands. 'Every day we face problems for which there are no off-the-shelf solution. Farmers must be fleet-of-foot, swift to adapt. They'd make very good politicians.' A couple of hours later, before I left, he returned to the subject. 'I don't drive a tractor now and I don't look after animals, so what am I? I'm an investment banker, trader, soil scientist, plant physiologist. If you're a farmer, you do all these things on a daily basis.'

Growing vegetables in an intensely competitive world requires high levels of skill, innovation and business acumen. 'When it comes to making decisions, you're talking plus or minus a week with cereals,' he said. 'If you sow or harvest a week late, it doesn't really matter. But with vegetables, you're talking about a twelve to twenty-four-hour window to get things right, and for salads just a few hours. You have to be very sharp and on the ball.'

Approximately half of the 22,500-acre estate is productive farmland, with a full-time labour force of thirty-seven people. Elveden is well known for its root crops, which include onions – it produces 6 per cent of our national output – potatoes, carrots and parsnips. Cereals are grown as a break crop to help control weeds and diseases and the estate also has a flock of sheep and a herd of suckler cattle. Most of the rest of the estate is devoted to conservation and forestry and includes some 4,000 acres of breckland, which supports an unusual flora and fauna, including rare species such as stone curlew, nightjar and woodlark. I had never been to the Brecks before and the landscape was unlike any I had come across. There were large blocks of pines, and many avenues of pine trees along the roadsides. With their dark green foliage and tall trunks, orange in the early morning sunlight, they made me think of the Scottish Highlands, but the countryside here was very flat, with a patchwork – part wild, part productive – of heathlands, which made me think of Dorset, and large fields, many with imposing irrigation gear that looked like giant insects on wheels. A peculiar hybrid, in other words, but beautiful in a strange way.

Elveden applied to join LEAF in 2000 as the estate wanted to supply Waitrose, which insisted on LEAF accreditation. Andrew said he appreciated the online auditing system developed by LEAF as it helps farmers to develop a 'to do' list. 'LEAF makes you consider the impact you are having, not just in the field, but beyond. For example, if we are using finite resources, could we substitute them with something else? LEAF has made us think differently about the way we operate.'

This was the first farm I had visited where there was a strong focus on research, on testing out different methods of production. Elveden conducts up to forty crop trials each year on the same block of land. For the past sixteen years, its own staff and a team of agronomists have carried out organic manure trials, looking at the impact of farmyard manure, pig manure, poultry manure and household waste on crop behaviour. They have also tested different varieties of crops under different conditions and studied their response to different irrigation regimes and fertilisers. 'Senior management has been very supportive of us trialling new ideas and technologies, and I have a free rein to be creative and innovative – although everything has to make financial sense in the end,' said Andrew. The trials have given him the confidence to grow new varieties and change the way crops are managed.

Elveden is one of thirty-eight LEAF demonstration farms and every year between 1,500 and 2,000 people come in groups to learn about its farming operations. These include college students, schoolchildren, farmers and others involved in the agricultural industry. The estate has gained a reputation for its wise use of water resources, and many visitors come to see how it operates its irrigation systems.

Andrew is a great advocate of promoting farming to a wide audience. 'I sit on boards related to conservation and land use, often with people who come from organisations that are critical of farmers,' he said. 'I think it's very important to interact with critics and explain how food production really works.' I liked his optimism. I don't know whether he voted for or against Brexit, but he seemed bullish about the future. 'There's too much talking Brexit down,' he said. 'Most of our seed comes from Denmark and France and we also buy a lot of our machinery from the continent, so a weaker pound is making these inputs more expensive. But we should see Brexit as a great opportunity to promote home-grown food production.'

While I was travelling round the country, good news about farming seemed to be in short supply in the national press. There was plenty about the travails of the dairy industry, low prices for many agricultural commodities and the uncertainty caused by Brexit, but little about our great farming successes. Yet there are plenty of these. One organisation with a good story to tell is the Produce World Group, which is now the largest supplier of fresh vegetables in the country, growing some 30,000 acres of vegetables on its own farms, in joint ventures and with dedicated grower groups. It also has its own organic grower, Taylorgrown, on the Houghton Estate, a lovely spread of rolling, well-wooded land in North Norfolk. This was where I met Andrew Burgess, whose family own Produce World, and Taylorgrown's general manager, Joe Rolfe, one fine July afternoon.

Andrew Burgess in a field of organic carrots.

In 1898, Andrew Burgess's great-grandfather lost an arm working for the London Brick Company near Peterborough. 'With the compensation money he received, he bought a field of potatoes, which he sold to markets in London,' said Andrew. 'Our family has been doing that ever since – supplying vegetables to market stalls and grocers, and now to supermarkets.'

After Andrew left college in 1986 he spent some time in Australia before returning to work on the family farm at Yaxley, a village a few miles south of Peterborough. 'When I got back, I set up a little shed with old-fashioned carrot-washing equipment, while my cousin Henry ran the potato side of the business,' he recalled. In 1984, a Waitrose buyer came to look at Henry's potatoes. He also took a liking to Andrew's carrots and ordered 100 boxes. In 1987, the potato and carrot business at Yaxley had a £1 million turnover. Eight years later, this had grown to £25 million. 'By then, we were no longer just growing vegetables at the home farm in Yaxley, but renting land elsewhere,' recalled Andrew. 'We hired a professional business expert and he said he would help us to increase our turnover to £100 million in five years. That's exactly what happened.'

As we wandered around the fields in late afternoon – first we went to look at the carrots, and then some brassicas – we talked about LEAF and the organic side of the business. 'I am very keen on both LEAF and organic,' said Andrew. 'You get some of the most progressive thinking in the organic world, and I like LEAF because it takes the best from both organic and conventional systems of farming.'

Andrew first decided to give the organic business a shot, as he put it, when he could see the growth in sales. 'I used to think it would be frightening, growing crops without chemicals, and when we began converting our first block of land to organic at Yaxley in 1997 we were right out of our comfort zone.' But they soon got the hang of it. By 2000, they were producing 5 tonnes of organic carrots a week; now Produce World sells 450 tonnes a week, with 90 per cent coming from their own land and three other growers. Produce World is now the sole supplier of leeks to one of our largest supermarkets, and accounts for 85 per cent of its onions as well. It is also the largest supplier of organic carrots in the country.

The success of the organic enterprise depends on a number of key elements, including long-term rotations and good soil health, weeding by hand and with camera-steered mechanical hoers, and working with researchers from Stockbridge Technology Centre to encourage predatory insects. 'Long before it became in vogue, we were establishing beetle banks and field margins to encourage ladybirds, lacewings and other insects which eat aphids, the main pest on carrots,' explained Andrew. 'We are making as much use of science as we can.' Carrot yields on the organic farms are now almost as impressive as they are on conventional farms, which wasn't something that Andrew expected.

At Houghton Hall, Produce World has also benefited from its collaboration with two other enterprises – one rearing organic pigs, the other organic poultry – from which it gets a healthy supply of manure. Pigs are also part of the rotation. 'We tend to follow brassicas with carrots, then spring cereals, then pigs, before putting down a three-year grass ley with clover, then cereals, before going back to brassicas again,' explained Joe Rolfe. Good soil health is one of the reasons why the estate doesn't suffer from black grass, a weed which has had a major impact on cereal yields on many arable farms.

I asked Andrew and Joe whether they were concerned about the impact that Brexit could have on hiring European labour. 'Yes,' said Andrew bluntly. Produce World employs some 500 staff, over 300 of whom come from Eastern Europe. 'It has been an absolute privilege working with Eastern Europeans. I admire their ethics, their family values. They work very hard and they are always cheerful.'

The company had done its best to find local labour for manual jobs, but with little success. A few years ago, the Burgesses took on thirty young men and women at the Home Farm in Yaxley. By the end of the first week, all but one had left. Why, I asked. 'Oh, you name it,' replied Andrew. 'It was too cold, too wet, the work was too repetitive, they didn't like being told what to do. There weren't enough tea breaks. So we replaced them with Eastern Europeans. Most are very well educated and some are now in managerial positions. If we sent all the Eastern Europeans home, we would starve in this country.'

Many of the company's Eastern European workers were very upset by the vote to leave the EU; they felt as if they'd been slapped in the face. 'One of my biggest jobs now is to make sure they feel secure again,' said Andrew. It is not only the horticultural sector that would be affected if Eastern European migrant labour was no longer available. Abattoirs, packing companies, milking parlours, processors and pig and poultry farms would also suffer.

Before I set off on my journey to see vegetable farmers, I spent some time with George Dunn, chief executive of the Tenant Farmers Association. Among other things, we talked about the relationship between supermarkets and farmers. He was impressed by Morrisons' recent venture designed to help dairy farmers. Customers buying its 'Milk for Farmers' brand were paying an extra 23p on a 4-pint carton. This went directly to farmers, in addition to what they were already being paid. Around 16 per cent of customers were choosing to buy this brand, a sign of willing-ness among some consumers to pay more for their food. On the other hand, said George, many farmers were still getting a raw deal from supermarkets, especially those involved in the horticul-tural sector. 'A supermarket might ring up a grower saying they want 1,000 head of lettuce at 25p a head on a Thursday morning,' explained George. 'So the grower gets all his pickers ready at 6 o'clock in the morning and just before they begin work he gets a call from the supermarket saying they only want 500 head of let-tuce, and they'll only pay 20p each. If the grower doesn't go along with it, the supermarket will buy elsewhere. That happens a lot.'

While farmers have become increasingly efficient, doubling yields of most vegetables over the past thirty years, the grower's share of the profits has steadily declined. 'Virtually everything we grow is much cheaper to buy now than it used to be,' said Andrew Burgess. 'When I left college in 1986, the wholesale price of carrots was around 21p a pound. This winter it was 13p a pound. In the mid-1980s you could buy a bit of land, grow carrots on it and pay back your loan within a year. Now you would be working half a lifetime to pay off your debt.'

The food retail sector has undergone dramatic changes since 2008. The financial crisis encouraged what Andrew described as 'a rush to discount', with two German-owned supermarkets, Lidl and Aldi, disrupting the rules of the marketplace. 'They've been very clever,' said Andrew. 'Unlike companies like Sainsbury's, who stock over 20,000 lines, they have just 2,000 or so lines and they concentrate on those. To get people through the door, they "invest" in five or six items people always want to buy, things like bread, baked beans, potatoes, carrots. Discount supermarkets often pay suppliers more than they charge their customers for these products.' This has led to a price war, with the big supermarkets – Marks & Spencer is not included in this bracket – pushing down prices in order to compete. All of this has put massive pressure on suppliers. 'Prices for farmers never go up in the long term; they always go down,' said Andrew. 'Supermarkets often ask us to give them our best price. You put your quote in, then they come back to you and say, "that's too much". So you then end up having to haggle.'

When my parents were born, in the early years of the twentieth century, the average household spent around 50 per cent of its disposable income on food. By the time I arrived in the 1950s, the food bill still accounted for one-third of average incomes. Today the equivalent figure is 10 per cent. So for every pound the average family earns, just 10p goes on the most important thing in the world: food – and that's not counting trips to McDonald's or The Ivy. Surveys of people using supermarkets invariably find that they want their food to be sustainably produced, with high standards of animal welfare. But once they start loading up their trolleys, most people are guided by cheapness rather than morality. I suspect that most shoppers have little idea about the efforts made by farmers and growers to meet increasingly rigorous standards, in terms of methods of production, food safety and welfare. Complying with these standards comes at a cost to the producer, a cost which many consumers seem unwilling to share.

I had planned to end my journey through the world of vegetables on the Houghton Estate, but I was intrigued by the symbiotic relationship between three separate organic enterprises – vegetables, pigs and poultry – and decided to call in to see the founder and joint managing director of Traditional Norfolk Poultry (TNP), which sources organic chicken from Houghton, when I was staying near its headquarters in Thetford in early August.

When he left agricultural college some twenty-eight years ago, Mark Gorton realised that it was going to be hard work getting into farming if he didn't want to work as an employee. He didn't have the capital to buy a farm of his own, but he was fortunate in that his parents had bought a house in the country with a couple of acres of land. When he was a teenager, he reared pigs and grew

vegetables to earn pocket money. After he left college, he and
a friend, now joint managing director of TNP, decided to raise
turkeys for Christmas at his parents' house. They started with
twelve birds, which they sold to friends. Gradually they built up
the business, moving into chickens and supplying local butchers.
Today, they have 1 million chickens at any one time on around
fifty farms, and their free-range and organic birds are sold by most
major supermarkets.

Mark is justifiably proud of the welfare standards, and taste, of
his chickens – TNP has even developed a new breed, the Norfolk
Black – but he didn't criticise the intensive broiler industry, which
supplies the vast majority of the 24 million chickens we consume
in the UK every week. There is, nevertheless, a great difference
between free-range birds and broiler birds. Ferrari broiler chickens, as Mark called them, are fed a high-octane diet and killed
after just thirty-two days, when they weigh around 2 kilograms;
the broiler industry is all about producing meat as rapidly and
efficiently as possible. TNP's free-range and organic chickens, in
contrast, are slaughtered at between fifty-six and seventy-seven
days and they are fed a varied diet, some of which they glean in
the open air. One person can look after 250,000 broiler birds, but
on TNP's farms it takes two or three people – mainly Eastern
Europeans – to look after 40,000 chickens.

Mark took me to see two farms owned by the company in
Thetford Forest. There was an end of summer feel about the countryside, with rose bay willow herb lining the hedgerows and flocks
of goldfinches feeding on teasels and other seed-bearing plants.

One of the farms we visited had 15,000 chickens, with around 1,000 birds in each of the fifteen mobile sheds. They had access not just to open pasture, but to birch and oak woodland, where some were searching for insects, seeds and other wild food. This contributes to their excellent taste. Once a batch of chickens has been harvested and sent for processing at TNP's headquarters, the sheds are dragged by tractor to another site, where they are washed and disinfected before being restocked. On the Houghton Estate, the chicken manure is used by Produce World on its vegetable fields. The manure from the farms I visited in Thetford Forest is given free of charge to local farmers.

Mark seemed a relaxed and personable character, but when I arrived at his offices he was clearly agitated. He told me that he'd been looking at a website for high-class barbecues on which celebrity chefs provide recipes and commentary on food issues. 'I just read this piece by a chef saying you can't buy quality meat from supermarkets, that you should always shop at your local butcher,' explained Mark. 'That is absolute rubbish! Almost everybody seems to assume that if you buy from the local butcher, the meat must have been locally sourced. But that's often not the case. Anyway, just because something is local doesn't mean it's good – it could come from one of the worst farms in the world!'

He returned to the subject when we were looking around his chickens. Businesses like TNP, he explained, are subject to intense scrutiny from supermarkets, certification bodies and government ministries. 'If we don't achieve certain standards, we simply couldn't sell to supermarkets,' said Mark. At a very minimum, supermarkets require Red Tractor assurance, which guarantees that producers meet certain standards. 'I don't think people give the industry the credit it deserves for its welfare standards,' said Mark. 'And I'm not just talking about free-range businesses like ours.'

I asked Mark how I would recognise his chickens in the supermarket. He explained that the particular chain where we always buy our free-range chickens doesn't put the provenance of chickens on the labels, just a licence number: 5450 in his case. The next weekend, I bought one of his chickens for just over £9. It provided a Sunday roast for three, with sufficient leftovers to make a stir-fry for two the following evening. From the carcass we made stock for a leek soup, which kept two of us going for two lunches. It is true that I could have bought three bottom-of-the-range broiler chickens for the price of one of Mark's, which implies that his birds are expensive. There is another way of looking at it: £3 or so for a chicken of any sort is absurdly cheap; and £10 for a top-quality free-range chicken is still cheap.

Before I set off on my travels in Yorkshire, Harry Marwood and I spent a day on the farm where we used to work in the early 1970s. Much had changed since those days. It used to be a mixed farm with two herds of dairy cows; and, before my time, there were beef cattle and sheep as well. A good portion of land was devoted to providing the feed to sustain the cows – grass in summer, silage and some cereals in winter – and the cows provided a plentiful supply of slurry which was applied to the land growing arable crops. The only livestock we saw this time was a group of thirty or so beef cattle being fattened indoors on barley.

This is a typical story, part of the ever increasing trend towards specialisation. In many parts of the country, mixed farming has almost disappeared, despite the fact that its inherent nature – the recycling of animal waste and crop residues – means it is one of the most sustainable ways of using the land. Although Produce World's enterprise at Houghton was not, strictly speaking, a mixed farm, it was drawing on all the virtues of mixed farming by collaborating with two livestock businesses. Getting fertility out of the back end of an animal makes more environmental sense than making it from non-renewable resources like oil. The old-fashioned ways are sometimes the best.

6 Outdoors Good, Indoors Bad?

I n *Lark Rise to Candleford*, Flora Thompson provided a vivid account of village life in Oxfordshire towards the end of the nineteenth century. Every cottage in Lark Rise had a lean-to pigsty. 'During its lifetime the pig was an important member of the family, and its health and condition were regularly reported in letters to children away from home, together with news of their brothers and sisters,' she wrote. 'Men callers on Sunday afternoons came, not to see the family, but the pig, and would lounge with its owner against the pigsty door for an hour, scratching piggy's back and praising his points or turning up their own noses in criticism.'

A good pig fattening in the sty promised a good winter. But it wasn't just rural workers who relied on the family pig as a source of protein. One of my first jobs after leaving university involved interviewing retired miners, many of them in their eighties, in the Durham coalfields. Certain themes recurred in all my conversations: mining tragedies, such as the West Stanley Colliery explosion; the General Strike of 1926; the joys of rabbiting with whippets; and the importance of the family pig. The miners would recount how the pigs were fed and kept, and the manner of their killing, which involved a metal punch being hammered into their brains before their throats were slit and the blood collected to make black pudding.

You may recall from the previous chapter that an old farming friend, Harry Marwood, introduced me to potato farmers in Yorkshire. Before we went to see them, Harry and I visited our old haunts in Wensleydale and he took me to see the farm where he had been born, near the village of Crakehall. He was the youngest of ten children and could recall his upbringing with remarkable clarity. Cobshaw Farm, which the Marwoods rented, covered 184 acres and was typical of mixed farms of that period, with cereals, root crops, grass, sheep, dairy cows, beef cattle, pigs and assorted poultry.

From the age of five, Harry would get up at 6 o'clock in the morning, put on his work clothes and feed the pigs and calves. Then he changed into school clothes, had breakfast, walked or cycled 2½ miles to school, returned the same way in the afternoon, put his work clothes back on, fed the pigs and calves, then helped his father prepare the rations for the next day, holding a paraffin lamp in winter as the farm had no electricity.

The family had a dozen or so sows – predominantly Large Whites, although Harry's father also bought the odd Saddleback, Tamworth and Berkshire if he got what he considered a bargain at an auction – and these slept in individual tin sheds in a small paddock. Piglets were weaned at eight weeks – methods of production were less intensive in those days – rather than four weeks or less, which is what happens in most piggeries nowadays. They were then finished, or fattened, indoors on straw. Most of the fat pigs would go to market, but the family always killed one or two each year for home consumption. 'In those days, every locality would have a pig killer and he would come to the farm when we needed him,' recalled Harry. 'Whoever was unlucky that day would have to hold a punch to the pig's head, and the pig killer would hit it with a mallet.'

'Did they not struggle?' I asked.

'Not much. There was a noose round the upper jaw, so they were tethered, and the pigs were very docile. They'd been hand fed, and there was probably more fat on them than lean.'

As soon as the pig fell to the ground, its throat was slit and it was de-bristled with scalding water. Then it was hung up for twenty-four hours for the flesh to stiffen before the pig killer returned to dismember the carcass. Harry's father cured the hams and bacon on a lead-lined table, using salt and saltpetre, and the preserved meats were hung from the kitchen ceiling and consumed over the coming months. 'Some things we had to eat fresh, like the livers and kidneys, so we would feast on offal for a few days after the pig was killed, and bits like the nostrils and cheeks were made into a meatloaf. As they used to say: the only bit you didn't eat was the pig's squeal.'

When Harry was growing up in the 1950s, one farm in every three had pigs. Now it is fewer than one farm in every 150, and the average number of pigs per unit – the language of pig production reflects its industrial nature nowadays – has risen from around seventy to over 6,000. There are some 10,000 pig farms in the UK, with 92 per cent of production coming from just 1,600 farms, including ten companies which account for 35 per cent of the breeding herd. The key elements of modern pig production are genetic selection for fast-growing, lean animals and maximum production of piglets, disease control, which sometimes requires the use of antibiotics, and efficient feed conversion. No other animal, apart from the broiler chicken, has been subject to such a degree of intensification. In the UK, approximately 40 per cent of our breeding sows live outdoors, but the vast majority of their progeny are finished indoors, some on straw yards, the majority on concrete or plastic slats in buildings that are artificially lit and heated.

I decided to focus in this book on the welfare of pigs, rather than chickens, which have been subject to similar degrees of industrialisation, for a very simple reason. All the evidence suggests that pigs possess a greater degree of intelligence than chickens, and possibly most other livestock. We humans have developed a close relationship with pigs, not dissimilar to the one we have with dogs. Indeed, pigs are said to make good pets; when I was working in the Durham coalfields I met a farmer who had a pig which spent time in his living room. In short, I find it easier to empathise with a pig than I do with a hen and consequently its welfare issues seem more pressing and interesting.

Many critics of factory farming begin from the premise that it is morally wrong to keep animals in close confinement. Indeed, I have done so myself. Some thirty years ago, I wrote a book with Richard D. North called *Working the Land: A New Plan for a Healthy*

Agriculture. I was responsible for the research and writing, while Richard, whose earlier book *The Animals Report* had explored issues related to farm animal intensification, brought a critical eye to the project. I have also written about outdoor pig farming for the Soil Association, more about which shortly. These projects involved plenty of fieldwork, all of which focused on farms that were doing things beautifully – as I saw it. I never visited intensive piggeries or broiler chicken houses and when it came to discussing animal welfare, I simply assumed that rearing livestock outdoors is inherently better than rearing them indoors in confined spaces, and marshalled the facts to prove the point. This time, I was determined to approach the subject with a more open mind.

The first pig farmer I went to see was Rob Shepherd. Unlike many farmers, Rob has a serious hinterland. He served in the Light Infantry after leaving school, studied agriculture at the Royal Agricultural College in Cirencester, travelled across the Sahara on a motorbike to West Africa, and served for many years as a reservist in the SAS following his return to the family farm on the chalk downs in Hampshire in the 1990s. He has a large herd of outdoor sows, approximately half of whose progeny are finished indoors in an intensive unit.

I called in to see Rob at his farm near Fordingbridge on a fine morning in mid-October. He looked much as you might expect somebody who had served in the SAS to look: fit, rugged and confident, and his manner of speaking matched his appearance. Before looking round, we sat in his office, drinking strong coffee and talking about the volatile nature of pig farming. 'You have to remember that this is a global business,' said Rob. 'There's one guy in the United States who owns more sows than the whole of the UK, and half the world's pigs are in China. We are just a tiny speck in the world pig production chain, and we live and die by what happens globally.' Rob was in buoyant mood; the Brexit vote and subsequent fall in the value of the pound had done wonders for his business during the past few months. Like many other pig farmers, he was benefiting from the rise in demand for UK pork meat in China, where consumers have become increasingly concerned about the safety of local food production.

Before the EU referendum in June 2016, Rob was getting about £60,000 for the 800 or so pigs he was sending to market each week. This was at least £6,000 less than the cost of production, which meant that he was losing a minimum of £24,000 a month. It had been even worse earlier in the year, but his fortunes had recently changed for the better. 'In February, I sold one load of twenty-week-old pigs for £67 a head. This October I sold pigs which were exactly the same specification for £115 a head. That's typical of the pig business. There have never been any direct subsidies for pigs and everything depends on supply and demand and the currency market. If cereals are down in price – feed accounts for about three-quarters of our costs – and the demand for pork is high, then you can make good money. But when the price of feed is high, the opposite can happen.'

Rob grows around 1,500 acres of combinable crops, such as wheat, barley and maize, and the rest of the land he farms is down to permanent pasture or grass leys. He used to have his own herd of beef cattle, but that became unprofitable and he now agists – looks after and feeds – cattle for other farmers, as well as a flock of sheep. His principal livestock business, however, is pigs. He has 2,000 outdoor sows and some 8,000 pigs in the finishing unit at any one time.

Approximately half the pigs which he sells come out of the indoor unit. They are either 'heavy cutters' – bog-standard super-market pigs without a premium, as Rob put it – which are sold for slaughter at twenty weeks old; or younger pigs, weighing some 40 kilograms, which are sold to farmers who will take them up to their slaughtering weight. All these pigs are Red Tractor Assured; in other words, they have been raised under systems of production which are regularly inspected by vets and which guarantee certain standards in terms of welfare, hygiene, health, the state of the buildings and so forth. In the UK, none of our major supermarkets will sell livestock products which come from systems which do not comply, at the very minimum, with Red Tractor Assurance.

Rob's outdoor sows, on the other hand, are RSPCA Assured, which means they are kept in conditions which meet the RSPCA's welfare standards. He sells around half his piglets as four-week-old weaners to farmers who will finish them up in straw yards to standards approved by the RSPCA. Pork from these animals will be labelled as such and fetch a higher price than the pork coming out of more intensive units. This is how the RSPCA puts it: 'The

RSPCA Assured label makes it easy to recognise products from animals that have had a better life, so you can feel good about your choice when shopping and eating out.'

'Pig heaven, wouldn't you say?' asked Rob when he took me to see the outdoor sows on the hillside above his office and home. And indeed, it did look very lovely. The pigs had recently been moved here, having spent a year in another block of fields, and they hadn't had time to root up all the grass. Each of the sows that were farrowing – giving birth – had its own ark and the pregnant sows were in groups in larger pens. This is ideal country for outdoor pigs, the chalk downs having light, free-draining soils, and all looked well under the hazy autumn sun. It was very much my idea of how a pig farm should look.

Rob Shepherd with one of his outdoor sows.

The majority of farmers who have opted for outdoor farrowing systems have done so for reasons of affordability. It costs around £250,000 to set up a 1,000-sow outdoor unit; ten times that much to build a similar-sized indoor unit. However, being outdoors doesn't guarantee high standards of welfare, according to Rob. 'Last year, it started raining here in August and it carried on raining all through the winter. The conditions were awful for the pigs, even worse for my staff. When the pigs were in the arks, they were okay, but outside they were knee-deep in mud for months on end. With outdoor pigs, you feed them on the ground, so the food was all mixed up with mud and pig shit – that's not good for their health. When conditions are like that, I think pigs are better off indoors.'

There are other problems with outdoor piggeries too. Rob's farm has a large population of foxes, badgers and ravens, predators which will kill young piglets when they have the opportunity. He couldn't put a figure to how many he loses, and indeed I haven't been able to find figures for the impact of predation on outdoor piglets. Likewise, although Rob puts heavy plastic curtains over the ark doors in winter, there is still a significant risk of newborn piglets dying when it is cold. Despite these problems, his pre-weaning mortality rate is less than 10 per cent, which is low for an outdoor unit.

'To be honest, I can't see much of a future with outdoor pigs,' he said as we made down the hill towards his finishing unit. 'If the winters get wetter with climate change, how am I going to find staff who will be willing to do the job? I suppose we will be able to get migrant labour for a while, but even that's not certain.'

Of course, a lot of people spend their lives working outdoors, but working in muddy fields with livestock in winter can be a particularly tough and dispiriting experience.

In recent years, many outdoor piggeries have gone bust. The largest outdoor herds are now mostly owned by abattoirs, processors and supermarkets, rather than family farmers. They need products that can be labelled 'outdoor bred' or 'outdoor reared' to satisfy consumers who believe that outdoor units provide better welfare than their indoor counterparts. Outdoor bred means the piglets have been kept outdoors until weaned, generally at four weeks, before being finished indoors. Outdoor reared is a somewhat looser term, implying that the pigs have been kept outdoors for approximately half their life. In neither case does it mean that pigs have been kept outdoors throughout their entire life, and I wonder how many consumers realise this.

Having never been inside a modern indoor piggery, I wasn't sure what to expect. We began with the recently weaned piglets, which were in batches of around forty in large wooden boxes with a lid and a heat lamp, with access to a small slatted area where they could do their business. They looked like glistening pink torpedoes. Rob then showed me the various stages in the pigs' maturation. One group were on straw in a large yard, but most were on concrete and slats in long sheds. In a slatted system like this, the pigs' faeces and urine fall into a slurry collector, and the slurry is used to fertilise the farm's arable land.

'Could I bring your friends from London here?' asked Rob when we reached the final building, which housed pigs that would soon be ready for slaughter. 'Or do you think this is the unacceptable face of modern farming?'

'Well,' I replied after some thought, 'I imagine they would be happy to see the pigs on straw, but probably not on slats.' And on reflection, this is pretty much how I felt. I preferred the sight of Rob's pigs sniffing around on straw to the last batch we saw, which were being finished on concrete and slats. Why? Largely because they looked more comfortable, more at ease with their environment. I accept that this is a wholly subjective judgement.

Although modern pigs bear little resemblance to their omnivorous, forest-dwelling ancestors, being short-snouted, pink and almost hairless, they have retained many of their behavioural traits. They like to live in small groups of around ten mature females and these groups have strong social bonds. Given the chance, they will spend much of their time foraging and rooting and they are highly inquisitive. Pigs are fastidious and they will do their best not to urinate and defecate in the areas where they lie.

Some twenty-five years ago, as you may remember from Chapter 1, the UK Farm Animal Welfare Council developed the 'Five Freedoms' as a guide to what livestock keepers should aim for. These are freedom from thirst, hunger and malnutrition; freedom from discomfort; freedom from pain, injury and disease; freedom from fear and distress; and freedom to express normal behaviour. Keeping pigs on concrete and slats clearly denies the animals the last of these five freedoms, as they are unable to perform many of the activities they enjoy, such as rooting and foraging and, in the case of farrowing sows, making a nest out of straw.

One of my field trips for *Batteries Not Included*, a report on organic farming and animal welfare commissioned by the Soil Association, took me to an organic pig farm in the Cotswolds. On a sunny summer's day, Sam and Helen Wade showed me round Eastleach Downs Farm and told me about their trajectory from working in intensive pig systems through to managing their own organic farm. Between leaving a company which specialised in creating commercial breeding stock and establishing their own farm, they managed a large outdoor unit in Hampshire. Over the years, they explained, they became increasingly concerned about the welfare of the pigs they were managing. They felt the piglets were being weaned too early, at three weeks, and they began to ask themselves, in Helen's words: 'What's the point of doing this, of rearing pigs outdoors, and then sending them into an intensive system with lots of welfare problems?'

Organisations like the RSPCA and Soil Association are critical of a number of practices used in intensive pig units. These include tail docking, which is designed to reduce the risk of tail biting in older pigs, and teeth clipping, the purpose of which is to reduce damage to sows' teats during suckling. According to welfare organisations, both these practices, or 'mutilations', can cause significant pain to piglets. Indeed, the EU stipulates that they should only be used as a measure of last resort. One practice that is used with some outdoor sows is nose ringing, which is done to reduce rooting of the ground. Under organic systems, none of these practices are allowed.

Other key concerns for animal welfare organisations relate to the use of farrowing crates and the conditions under which many pigs are finished, on hard slatted floors. Typically, sows are placed in farrowing crates – metal cages with a concrete, partially slatted floor – shortly before the piglets are born. The main function of farrowing crates is to reduce piglet mortality: metal bars keep the sows out of the area where the piglets can lie, often on a heated pad. The sows remain in the crate until the piglets are weaned at around four weeks. Critics of the system say that close confinement can cause muscle weakness and swelling of the joints. The sows may also exhibit behaviour which indicates high levels of stress or boredom, such as biting the bars of the crates and shaking their heads. Farrowing crates have been banned in Sweden, Norway and Switzerland but they are still widely used in the UK and many other parts of the world.

As far as finishing pigs on hard, slatted floors is concerned, critics say these frequently cause foot problems and prevent pigs from performing their natural behaviour, such as rooting and exploring the environment. As a result, they are likely to become bored and frustrated, and this can lead to fighting and injury. This is why tail docking is carried out in many indoor systems.

When I met the Wades, some fifteen years ago, it was obvious that a great deal of thought had gone into creating a welfare-friendly system. Although many organic farmers favour traditional

A native breed sow and her piglets.

breeds such as Saddlebacks, Tamworths and Gloucester Old Spots, the Wades had opted for a cross between a high-performance modern strain and a Duroc boar. This has less fat than traditional breeds, enough colour to resist sunburn, and performs well under organic conditions. The sows are kept in groups until they are ready to farrow. They are then transferred to individual arks, where they are provided with barley straw as bedding. Piglets are weaned at eight weeks and kept in family groups, outdoors, until they reach slaughtering weight.

You will pay more for organic pork and pork products reared outdoors than you will for pork produced in intensive indoor systems. Just to give you a comparison, at the time of writing Eastleach Downs Farm back bacon cost £15.50 a kilogram; the equivalent in Tesco cost £6.67. Comparable figures for sausages were £8.00 a kilogram and £3.31. A key question here is: are you paying more for higher standards of welfare, or is it possible to satisfy the lower end of the market, which is where most people shop, without seriously compromising animal welfare? In search of an answer, I visited two more farmers – one in Yorkshire, the other in Devon – both of whom manage relatively intensive systems of production. Unlike Rob Shepherd, they farrow all of their sows indoors and in crates, although one keeps them in confinement longer than the other.

I called in to see Peter Batty, who runs a large farming enterprise in a village near Pontefract, on a bitterly cold day in January. He gave the impression of a life lived energetically and to the full. His mobile phone rang almost constantly, which was unsurprising as he has his fingers in many different pies. He has 600 sows, kept indoors for all of their lives, and a modern finishing unit, which has cost him many millions of pounds to develop. He is also involved in the arable side of the family farm and grows potatoes and sugar beet as well as a large area of cereals, which provide his pig unit with around two-thirds of its feed.

The conversation began with Peter bemoaning the failure of the farming industry to explain the welfare implications of different types of piggery. 'Most consumers have a picture in their heads of sows dancing around in a daisy paddock when they think of outdoor pig production, even though the reality is often far from that,' he said. 'Imagine what it's like for pigs in outdoor units in Scotland today, with temperatures below freezing.' Modern breeds of pig, bred for their leanness, are particularly susceptible to the cold. Peter also pointed out that there are certain buzzwords – 'natural', 'outdoor', 'local' – which are associated in the public's mind with high standards of welfare. 'The flipside is "intensive", "indoor" – people think that must be bad, but it isn't and needn't be.'

In 1999, a year before Peter took over the running of the pig unit, the UK introduced legislation which banned the tethering of sows and the use of dry sow stalls. The tether consisted of a chain attached to a collar around the neck or a belt around the middle of the sow; sow stalls, which were also designed to restrict movement, consisted of metal crates with concrete, partially slatted floors. The ban on these practices was all to the good, said Peter: 'Tethers were horrendous, and they could cut deep into the pig's skin.' The use of dry sow stalls meant that breeding animals spent virtually their entire lives in a narrow crate in which they were unable to turn round or express any form of natural behaviour. The only exercise they got was walking a few yards from the dry sow stall to a farrowing crate shortly before they gave birth.

Before Peter took over the pig unit, his uncle had decided that with the growing emphasis on animal welfare it would be best to keep growing and finishing pigs on straw and have simple buildings with passages for cleaning out. And indeed, schemes managed by the RSPCA and organic certifiers like the Soil Association stipulate that pigs must be kept on straw, or outdoors, as they consider this best for their welfare. At the time of my visit, Peter's growing pigs spent at least some of their life on straw, before being finished on slats, but this was about to change.

From an economic point of view, using straw bedding is inefficient, as shifting straw involves a lot of labour. Peter also had concerns about straw from a welfare point of view. 'When the straw is fresh the pigs look as happy as Larry, but after several weeks you get an increase in disease problems.' In his most recently constructed building – he was expecting to spend a further £1 million on a modernisation programme this year – pigs are kept in groups of thirty or so on concrete and slats. He has total control of temperature and ventilation and as a result he says that the pigs are never too hot or too cold. At the time of my visit, he was thinking of moving away from straw altogether, apart from in the buildings where the pregnant sows are kept. He mentioned, in passing, that when his wife had seen his latest building, which contained half a dozen or so groups of pigs close to their slaughtering age, she told him that she preferred to see them on straw, rather than on concrete and slats as they were here. I could see why.

'If you want pigs to grow well, you need to create a stress-free environment,' explained Peter. His pigs are kept in their birth or weaner groups, so they all know each other and have an established social group. 'If I was to take one of these pigs out of this pen and put it in that pen', he said as he showed me round, 'the chances are they would kill it by tomorrow. You get a very high level of aggression with pigs, and we do everything we can to ensure there is no bullying.' Providing the pigs with all their needs –

in terms of food, water and a constant temperature around 20°C – also helps to reduce stress.

Shortly before the sows give birth, they are put into farrowing crates. Peter has four Portapig units – portacabins with two rows of ten crates either side of a central passage – and the sows remain here until the piglets are weaned at approximately four weeks old. As we have seen, animal welfare organisations believe these crates compromise a sow's welfare. They certainly ensure that she cannot perform her natural behaviour, such as creating a nest before farrowing, or indeed moving around or rooting. But the crates do help to reduce the risk of the piglets being squashed by their mothers. 'Our piglet mortality here is 7 per cent,' said Peter with obvious pride. 'That compares with around 15 per cent for outdoor pigs.' So from a piglet's point of view, if not from a sow's, farrowing crates have much to recommend them.

My final visit, in the early spring, was to the village of Clyst St Mary in Devon. Over the past twenty years, Andrew Freemantle, the owner of Kenniford Farms, has developed a thriving business finishing pigs for the market and processing some himself for direct sale to the public. Peter Batty had suggested I visit Andrew as he had interesting things to say on the subject of animal welfare and the use of farrowing crates for sows.

Every week, Andrew sends around 180 pigs to the abattoir, 140 of which are sold to major food processors through a large farmer-owned pig cooperative. The rest are returned to the farm to be butchered. You can buy fresh pork at the farm shop and

Pigs near slaughtering weight in an intensive finishing unit.

cooked dishes made from Andrew's pigs at four catering trailers based at South Molton, Cullompton, Newton Abbot and Liskeard. The farm also does hog roasts at agricultural shows and festivals.

When I made my way to Clyst St Mary in early March there was more than a hint of spring in the air, with snowdrops and daffodils on the village greens and the first splashes of white blossom on the blackthorn bushes. I arrived around mid-morning and Andrew suggested I should join him and his staff for coffee. 'It's the diligence of the staff that determines the standards of welfare in a pig unit – more than the system itself,' he said as he introduced me to Grant Morrow, who manages the pig production side of the business.

Andrew and Grant led me into a large barn, from one end of which the visiting public can see some of his sows with their piglets. 'We started off having a free access system, where the sows could move around with their piglets,' explained Andrew. This qualified the farm for the RSPCA's Freedom Foods label – the forerunner to RSPCA Assured – but they found that piglet mortality was very high. 'We were losing too many piglets – mortality was around 17 per cent – so we decided to switch to conventional farrowing crates.' The use of crates meant that fewer piglets were accidentally killed by the sows and piglet mortality immediately fell to 11 per cent. This means that Kenniford Farm has saved the lives of around 700 piglets a year by adopting a system which is considered unsatisfactory by animal welfare organisations.

After twelve days, the sows and piglets are moved into larger stalls on straw. Known as Solari pens, these provide the sows with more space to move around at a time when the piglets are large enough to avoid being squashed. 'It doesn't make sense in cold, economic terms, but we think it's a good compromise,' said Grant, who has been working with pigs since the 1970s. He estimates that it takes some twenty hours a week looking after the fifty-four Solari pens, whereas it would take just five hours with traditional farrowing crates.

I asked whether the staff at Kenniford Farm practised tail docking and teeth clipping of piglets. 'Yes,' replied Andrew, 'and we would have serious welfare problems if we didn't.' He describes Kenniford on his website as a 'high welfare Red Tractor approved farm' and operates according to its protocols, one of which requires pig farmers to get permission from a vet if they want to dock tails and clip teeth. In the vet's opinion, it must be in the best interest of the pigs. Although the dry sows spend most of their time indoors, they have access to a large paddock for ten months a year, but they use it reluctantly, according to Andrew. 'The only reason they go out is because we feed them there,' he said.

While I was in the West Country, I dropped in for a lengthy conversation about pig welfare with Roger Blowey, the Gloucestershire vet who featured in the chapter about the dairy industry. Although Roger is best known for his work with cows, he has had over thirty years of experience working with pigs. One of the herds he attends to lives exclusively outdoors. He also looks after several indoor units. He was ideally placed to give me an objective view – insofar as it is possible to be objective about a subject that mixes morality, welfare and economics – on the merits and defects of different methods of production.

Despite their often placid appearance, pigs can exhibit high levels of aggression, given the opportunity. This didn't matter when sows were kept in either dry sow stalls or farrowing crates, as they never came into contact with one another. However, following the 1999 ban on tethers and dry sow stalls, animals which had been kept in what amounted to solitary confinement were now managed in small herds in straw yards before going into farrowing crates, a practice that came with its own set of problems, according to Roger.

'I am not saying I was in favour of sow stalls and tethers, but I can see the welfare advantages and disadvantages of different systems,' he explained. 'When you introduce pregnant sows into a yard where other pregnant sows have already established their social hierarchy – this is what happens now – you get continuous disruption of the social status. You often find that when a new batch of sows is put into the yards, they spend the first twenty-four hours sitting in the muck, because they can't get near the straw bedding, which has been taken over by the more dominant and established pigs. They are also likely to suffer from vulval biting and bullying. The point I'm trying to make is that sows are aggressive, and that you will get welfare problems in any system.'

As far as outdoor systems were concerned, Roger confirmed much of what I had already heard from the farmers I had visited. On the positive side, outdoor systems allow pigs to root around, express themselves and exhibit the sort of behaviour you would expect of pigs in the wild. They didn't, as far as he could tell, suffer from boredom, which can be a major issue for indoor pigs. There was also less tail biting than there is with indoor pigs, which precluded the need for tail docking. In any case, he said, the sows in an outdoor unit would probably attack you if you tried to dock their piglets' tails.

However, there are also many welfare problems associated with outdoor piggeries, of which he had had much first-hand experience.

Although he is semi-retired, he still looks after an outdoor herd of pedigree Tamworth pigs. 'In 2010 we had the most terrible winter,' recalled Roger. 'The pipes were frozen solid so it was difficult to get water into the troughs, and sows were farrowing in temperatures as low as −14°C. Many of the piglets just froze to death that winter.'

Roger also believes you tend to get some specific disease problems in outdoor piggeries, not least because pigs eat their food off ground which is often covered with faeces. During hot summers, pigs can suffer from sunburn, and this means that the sows are less likely to stand for the boar, so getting them in pig becomes a trickier task. Then, of course, there are the problems of predation, which can be considerable. 'You've also got to think about the welfare of the stockmen,' said Roger. 'One of the reasons why so many people began with their pigs outdoors and then brought them indoors is because they couldn't face another winter outdoors, struggling with the wet and cold.'

Indoor units have their advantages, as well as their disadvantages. As far as farrowing crates were concerned, Roger said he didn't find them in any way distasteful. 'I know that more piglets will survive when the sows are in crates,' he said. 'And it's easy to cross-foster piglets if a sow hasn't enough milk when they are kept in farrowing crates. It's much more difficult to do that on an outdoor pig farm.' There are other advantages of being indoors as well: the pigs and piglets are not likely to suffer from heat stress or extreme cold. If farmers are using straw as bedding, that should be fine, he thought, as long as the straw is changed regularly. If it isn't, then it can lead to a greater incidence of disease.

In terms of the disadvantages of indoor piggeries, boredom was an obvious factor. The Red Tractor guidelines stipulate that pigs should be provided with materials that they can chew and destroy, such as wood and rope. However, I was struck by how barren the environment seemed in the intensive units which I visited. As for tail biting, it is much more common in intensive indoor units than it is in outdoor piggeries, an obvious manifestation of stress or boredom. This is why many indoor units practise tail docking. When I discussed the issue with Peter Batty, he said he didn't think tail docking caused any great trauma to the piglets. In any case, he said, wasn't it strange that so much fuss was made about tail docking of piglets but nobody seemed to object to the castration of young lambs, a practice which is carried out by putting an elastic band around the base of the scrotum?

I asked Roger about this. 'Yes, tail docking is a mutilation, and I think it's something which we haven't got to grips with properly in this country. However, if it's done quickly and cleanly with heated clippers or electric guillotines, there is no apparent adverse effect on the piglets. The pain outwardly exhibited by lambs when they are castrated appears to be much greater. Of course, you could say that all these practices are wrong.'

So what was his overall verdict about the relative merits of indoor and outdoor piggeries? There was no simple answer, he said. 'Sometimes pigs are better off inside. Sometimes pigs are better off outside. Sometimes pigs are better off on concrete slats. Sometimes pigs are better off on straw.' He added that standards of management are almost as important as the system itself.

A well-managed indoor piggery could have better welfare than a poorly managed outdoor piggery, and vice versa.

Although some hobby farmers and small-scale producers still raise pigs in a way that was familiar to anyone brought up in the 1950s and 1960s, the vast majority of animals now come from large, highly capitalised, relatively intensive operations. If you disapprove of this, for whatever reason, then you will need to either pay more for meat which comes from pigs produced to high welfare standards – although, as I have pointed out, outdoors does not necessarily mean better welfare than indoors – or consume beef and lamb produced in extensive, pasture-fed systems such as those described in chapters elsewhere in this book.

In terms of the welfare of farm animals, I think there has been good progress. When Richard D. North and I wrote *Working the Land* in the 1980s, two-thirds of the 750,000 sows in this country were kept throughout their pregnancies in dry sow stalls. As we have seen, these were banned in 1999, along with the practice of tethering sows. In the mid-1980s, around half the veal calves in the UK were reared in narrow crates in which they were unable to turn, and most were fed a diet lacking in roughage so they produced 'white veal'. Veal crates were banned in the UK in 1990, and in the EU in 2007. There have also been significant improvements in poultry welfare over recent years. From what I have seen on work assignments in Africa and Asia, free-range hens lead a better life than caged hens, and around half the eggs produced in this country now come from free-range hens. Our largest supermarket, Tesco, has pledged to stop selling eggs from caged hens by 2025, largely in response to a teenager's petition which attracted 280,000 signatures. Other supermarkets have already done so, or pledged to do so in future.

Much of the credit for these reforms should go to organisations like the RSPCA, Compassion in World Farming and the Soil Association. By arguing their case in a rational manner, and engaging with the farming industry, they have had a much greater influence than animal rights organisations such as Viva!, which would like to turn us all into vegans. In its own words: 'Viva! campaigns for a vegan world because most farmed animals spend their short and miserable lives in the filth of factory farms and are killed with sickening barbarity.' There is no denying that there are occasional episodes of animal abuse, for example in abattoirs – this is well documented – but to suggest that the vast majority of farm animals live miserable lives and are killed with sickening barbarity is hyperbolic to say the least.

I have a lot of time for someone like Jamie Oliver, who has taken a gradualist approach to influencing farm animal welfare. In 2015, he signed a multi-million-pound deal for frozen ready meals with a Brazilian company, Sadia, for which he received much criticism. Oliver responded as follows: 'If we want a big change, we need big corporations. Maybe I'm wrong. But I prefer to try. For me, being inside this machine, a company that is responsible for 18 per cent of the chicken in the world, is something positive.'

He believed that he had already made a difference to the welfare of over 40 million Brazilian chickens by convincing the company to provide perches for the birds and install systems to control the temperature.

Another well-known personality from the world of cookery, Jay Rayner, looked at the pros and cons of corporate agriculture in *A Greedy Man in a Hungry World* and came to the following conclusion. 'Given the challenges we face, large-scale agriculture makes an awful lot of sense. It just needs to be pursued by enlightened people.' When I was born, in the early 1950s, the world population was just over 2.5 billion; there are three times that many people today and the United Nations predicts there will be four times as many, 10 billion, by 2050. Feeding that many people from small family farms simply isn't an option.

Since the mid-1970s, average per capita consumption of pork meat in the UK has fallen by around 40 per cent, and the same goes for beef. In contrast, consumption of chicken meat – low-fat and easy to cook – has risen by 65 per cent. In these health-conscious times, with over one in every ten people in Britain being vegetarian, there is a good chance that consumption of red meat will fall further. Survival will be an increasingly tricky business for many farmers, even for those in the pig industry who have been used to getting by without state support.

'We get taught a lot of hard lessons with pigs, and nothing concentrates the mind like losing a quarter of a million pounds,' said Rob Shepherd as I was getting ready to leave his Hampshire farm. He believes that the best way of securing a viable future lies in diversification, and he has an interesting way of explaining this. 'People pay out of two pockets. Out of the first pocket they pay their food, energy bills, phone bills – and they'll do everything they can just to save 10p. Then there is the second pocket, a very deep pocket, for meals out, trips to Disney World and the cinema, new consumer goods – and they give much less thought to what these things cost. What I'm planning to do is get people to spend out of their second pocket rather than the first. We need to recognise that the land in the UK should be a bigger asset than just a farm production system.'

He was expecting direct subsidies to farmers, such as the Basic Payment Scheme, to end after 2020, and he was planning to invest whatever income he can make from his farming enterprises in non-farm activities, such as caravan storage, motocross events and music festivals. 'Those are the sort of things we're going to have to do if we're going to survive,' he said. 'I am at the top end of the age range of farmers who are thinking seriously about alternatives, but there are many progressive young farmers who know that they are going to have to use the land more imaginatively in future.'

7   Return of the Natives

Many millions of cattle, sheep, pigs and poultry are transported each year to British auction marts and abattoirs. Most people probably don't even notice. It is not exactly a hidden trade, but you need to get your eye in to distinguish between trucks that carry livestock and trucks delivering other goods. Until just a few generations ago the flow of livestock was there for all to see, with vast numbers of animals being walked, often over great distances, through villages and fields and along drovers' roads under the open sky.

When the novelist Daniel Defoe visited East Anglia in the early 1700s, he had much to say on the subject of cattle. In those days, according to Defoe, 'the gross of all the Scots cattle which come yearly into England' were brought to the village of St Faiths, now known as Horsham St Faith, a few miles north of Norwich, where they were sold to graziers who fattened them on the marshy land behind Yarmouth and Lowestoft. 'These Scots runts, so they call them, coming out of the cold and barren mountains of the Highlands in Scotland, feed so eagerly on the rich pasture in these marshes, that they thrive in an unusual manner, and grow monstrously fat,' wrote Defoe. The meat was so delicious that the English preferred Scottish beef to their own cattle, and once fattened, large numbers were taken from East Anglia to markets in and around London. The first cattle fair at St Faiths was held in the early twelfth century, the last in 1872. The site, to the west of the village, is now taken up by cereal fields sandwiched between Calf Lane and Bullock Hill.

The historian G. M. Trevelyan estimated that even before the Acts of Union in 1707, Scotland sent 30,000 head of cattle to England each year. When Defoe ventured to the far north of Scotland – 'this frightful country', as he called it – he noted that Caithness bred large quantities of black cattle which were sold in England. These would almost certainly have been the ancestors of today's Aberdeen Angus. Hardy, naturally hornless and of modest size, characteristics which made them easy to handle, Aberdeen Angus were sent not just to England for fattening, but in the nineteenth century to the United States, where they are now the most populous breed, and to Canada, Argentina, Australia and New Zealand. Native breeds of beef have been one of Scotland's greatest exports.

Over 50,000 farms in the UK have a herd of beef cattle and they make a major contribution to the rural economy. In 2015, for example, the UK beef and veal sector was worth £2.7 billion, compared to £2.2 billion for poultry and just over £1 billion each for pigs and sheep-meat. Our home-grown beef – we import approximately a quarter of our needs – comes from both the dairy herd and the suckler herd, with roughly the same quantity from each. While calves from the dairy herd are taken from their mothers at birth, those in the suckler herd stay with their mothers till they are weaned at six to eight months. The most intensive beef system involves fattening uncastrated males from the dairy herd, generally indoors, on a diet based on cereals. Suckler herds, in contrast, spend the summer outdoors, and in some cases the winter months as well. They spend most of their lives eating grass or grass products, although they may be given supplements such as cereals during the winter months and in the weeks prior to slaughter. Since the 1970s, lean, fast-maturing continental breeds, such as the Limousin, Charolais and Simmental, have played a dominant role in the beef industry, but native breeds like Aberdeen Angus, Hereford and Shorthorn are now become increasingly influential and they are providing some of the finest beef in the world.

In mid-September I headed to the Scottish Borders, an area the size of Yorkshire with a population of around 125,000 people, a quarter of whom make a living in agriculture. My first port of call was Morebattle Tofts Farm, whose owner, James Playfair-Hannay, is one of the leading breeders of Aberdeen Angus and Beef Shorthorn cattle. Seven generations ago, his relative James Playfair wrote the first document about Angus 'doddies', the forerunners of Aberdeen Angus, long before the first Angus herd book – a record of breeding pedigree – was published in 1862. Cattle are in the family's blood.

I had come to the Borders at a perfect time of year, the countryside at its fruitful and mellow best with the hedgerow bushes

— sloe, elder, hawthorn — spattered with purple and red berries and flocks of goldfinch and linnet feeding noisily on the seeds of wayside plants. About half the cereal crops had been harvested; the rest, golden still, awaited the combine harvester. A few stubble fields had already been ploughed and the upturned sods glinted in the late-summer sun like slabs of dark chocolate. There were gulls and rooks feeding on worms and insects, and wood pigeons filling their crops with spilt corn. On the approach to Kelso a large flock of lapwings wheeled, square-winged and jerkily, above the undulating landscape.

James was still out on the farm when I arrived, so I chatted to his wife Debbie while she prepared dinner. The kitchen was at the back of the farmhouse, which had a comfortable, lived-in feel to it. It was grand without being ostentatious, parts dating back many centuries. Debbie and I talked at first about mutual friends, Mike and Peta Keeble. Mike had been the manager of the Clifton Estate, in North Yorkshire, when I spent a year there as a farm student. I began researching this book shortly before Mike died in November 2015, and he suggested that I should explore the story of native breeds rather than continentals, and specifically Beef Shorthorns, whose virtues he had recently extolled in a pamphlet published by the Beef Shorthorn Society.

Soon after James returned, we sat down for dinner — tomato salad, beef stew, potatoes, green vegetables and plum crumble, all home produced — at the kitchen table. James reminded me of Mike Keeble: both are — were in Mike's case — fine talkers blessed with a great store of knowledge and a deep pool of anecdotes; both had thought long and hard about the state of the farming industry and its future. Over dinner, and later with a tumbler of Glenmorangie, James talked me through the history of Beef Shorthorns and his involvement with the breed.

Readers familiar with north-east England may well have drunk in a pub called the Durham Ox – there are many of them – and seen the signs outside which depict the beast in question. Born in 1796, the Durham Ox was a castrated Shorthorn of prodigious size and excellent conformation, bred by Charles Colling, who farmed with his brother Robert in the village of Barmpton near Darlington. For some five years the Durham Ox travelled round the country in a horse-drawn carriage to be shown at agricultural fairs and other events. A single day's admission fees in London in 1801 amounted to £97, a huge sum in those days.

A race of superior Shorthorn cattle had existed on the Yorkshire estates of the Dukes of Northumberland since the late 1500s, and from 1730 onwards various breeders, including the Colling brothers, began developing the cattle and recording their pedigree. In the nineteenth century, British farmers began exporting our native breeds and they helped to build the beef industry around the world. Shorthorns were the first to be sent across the Atlantic in large numbers, to be followed by Herefords – white-faced cattle descended from the small red cattle of Roman Britain and a larger Welsh breed – and Aberdeen Angus. Before long, the Americans and the Argentinians were sending large quantities of beef back to Britain and this had a profound influence on the nature of the beasts. 'In order to fit in the ships' chillers, the cattle had to be relatively small in size,' said James. 'Curious, if you think about it. You'd have thought they could have made the chillers bigger, rather than breed smaller cattle.'

When James left college in 1979, he returned home to small, unfashionable cows, the pedigree herd of Aberdeen Angus established by his grandfather just after the Second World War. However, James could see a future for native breeds and he decided to establish a herd of Shorthorns. Initially they were relatively small – a legacy of the transatlantic trade – but over the decades he and other farmers adopted breeding practices that increased their size. He has always taken great pleasure in showing his pedigree cattle – 'It's a good way to measure your achievements against your peers,' he said – and in 1991 one of his bulls was inter-breed champion at the Royal Highland Show.

James now runs about 100 Aberdeen Angus cows, fifty Shorthorn cows and what he described as a heap of cross-breeds on some 4,000 acres of land at Morebattle Tofts, Clifton-on-Bowmont and Yetholm Mains. 'We breed pure Aberdeen Angus and pure Shorthorns, but if they aren't top quality, we cross them,' he explained. For example, a second division Angus is put to a Shorthorn bull, and a second division Shorthorn to an Angus bull. This is a way of encouraging hybrid vigour in the suckler herd.

Next morning, after a breakfast of porridge oats and bacon, I accompanied James on his morning tour of the farm. We began our journey close to the farmhouse, where James provided some feed to half a dozen young bulls before dropping by to see fifteen older bulls which he was hoping to sell soon. They made a fine

One of James Playfair-Hannay's young Shorthorn bulls.

sight. One of the great attractions of Shorthorns is their colour, which can be red, red and white, roan, roan and white, and sometimes plain white. I asked whether any of these bulls had been put to a heifer yet. 'Yes, those three,' he said. 'They have been to the party and they have learnt how to dance.'

When James was growing up in the 1960s, most beef cattle came from dual-purpose dairy herds. Cows such as British Friesians, Dairy Shorthorns and Ayrshires were put to beef bulls like Herefords and Shorthorns. Their female progeny went into the dairy herd for milking; the males were fattened for beef. In the 1970s many dairy herds began using continental sires, such as Limousins and Simmentals, whose progeny could finish to heavier weights in a much shorter period of time than those of native breeds. Their meat was also leaner and less fatty. 'Native breeds began to go out of fashion, because they didn't put on kilos as rapidly and they had too much fat,' recalls James. 'When we joined the European Union, the health policy was all about reducing the fat content of meat.'

Times have changed and we are now witnessing a dramatic revival in the fortune of native breeds. In 2000, there were just 918 Shorthorn cattle registered with the Beef Shorthorn Society. Their number almost doubled by 2005, and 3,487 Beef Shorthorns were registered in 2016: a tiny figure when you consider that the UK beef breeding herd numbers some 1.5 million, but significant in terms of its rapid growth.

Morrisons, the fourth largest supermarket chain in the UK, now offers farmers a premium of 10p a kilogram, or around £75 a head, for Shorthorn cattle. When it launched the scheme in 2011 there weren't enough purebred animals in the country to meet its requirements, so the company gave a premium for cattle which had been sired by a Shorthorn bull. Now, thanks to the rapid increase in numbers, the premium is reserved for purebred Shorthorns. Most other supermarkets also offer a premium for native beef, and particularly Aberdeen Angus. James said he had recently bumped into a buyer from a supermarket who told him that since they had begun selling Aberdeen Angus as a 'top shelf product', complaints about their beef had fallen by 40 per cent.

Shorthorn farmers like James have a great story to tell. Not only are they producing succulent, tasty beef, they are making the best possible use of the available resources. 'Up here in the

Borders, we expect our cows to live on fresh air and grass – and produce high-quality calves,' he said. He reckons that less than 10 per cent of our beef farmers rear their animals on a diet of grass and grass-based forage. The rest, to a greater or lesser degree, follow the feeding system used in America's beef feedlots, where cattle are fattened on wheat, maize, other grains and protein-rich soya as well as grass-based forage.

The first-time heifers on James's farms are brought indoors for calving in the spring, so he can keep an eye on them, but subsequently they calve outside. 'There's no sitting up all night, waiting for cows to calve, like there is for many farmers,' he said. 'We might lose a few calves, but no more than the average.' This being one of the driest parts of Scotland, he is able to keep his cows outdoors throughout the winter. They eat dead grass – standing hay, as he calls it – and he provides them with small quantities of grain and a mineral block in the early months of the year before the grass starts growing again.

Almost half the wheat grown in Britain is now fed to animals, mostly to pigs and poultry, but also to dairy cows and beef cattle. Cereals are not part of their natural diet, so it is not surprising that this affects the quality and nutritional value of meat and milk. Converting grain to meat also comes with a high environmental cost. Take, for example, beef in feedlots, or concentrated animal feeding operations, as they are known in the US. According to one set of calculations, it requires over 15,000 litres of water to produce 1 kilogram of feedlot beef. This compares with 5,000 litres for 1 kilogram of cheese, 700 litres for 1 kilogram of apples, and

131 litres for 1 kilogram of carrots. One of the reasons why feed-lot beef is so water-hungry is because the average cow eats over 1,300 kilograms of grain in its lifetime, and grain production often requires a lot of water. These figures have been produced by WWF, which is a conservation body and not a research agency, so they should be treated with a degree of caution. But no matter how you do your calculations, the environmental footprint of grass-fed beef will always be smaller than that of more intensive production systems.

Spend time with beef farmers, especially those involved with native breeds, and there will be much talk about genetics and the science of breeding the best livestock for the market. You will hear about body condition scoring, conversion ratios, live weight gain, fertility, mothering qualities and milking ability. Farmers like James Playfair-Hannay have been steeped in beef genetics since an early age; others, like Spencer McCreery and his partner Arabella Harvey, are relatively new to the game.

Spencer and Arabella Harvey bought their first twelve Shorthorn cows at a sale of cattle in Carlisle in March 2015. The sale made headlines for all the wrong reasons, as inconsistencies in the birth dates registered by the breeders led to invalidation of the passports of seven cows. As passports are required by all cattle, this meant they could never be sold or moved again. I had read about this before I turned up one morning at Yester Farm,

Arabella Harvey and Spencer McCreery.

a handsome and well-wooded spread of land near the village of Gifford, to the east of Edinburgh, and I expected the subject to loom large in our conversation. After a while it did get a mention, but this feisty couple – Spencer was brought up on a farm in Northern Ireland, Arabella in the Scottish Borders – were far more eager to talk about the excitement of establishing a herd of Shorthorns than any of their misfortunes.

I had already eaten breakfast but Arabella, a warm and enthusiastic woman in her late thirties, insisted I have another. We ate scrambled eggs and cottage cheese on toast, washed down with strong coffee. The cheese, which was delicious and creamy without being unctuous, came from Yester Farm Dairies, which is owned by Spencer's brother and sister-in-law. While we were having breakfast, Spencer and Arabella explained how they had first established a herd of beef cattle some four years ago, fattening beef-cross calves from the dairy herd and cattle which they buy in the spring and sell in the autumn. More recently, they decided to develop another side to the business. 'We've always been interested in native breeds, and we liked the idea of seeing sucklers in the fields now the dairy cows are indoors all year round,' said Spencer.

A *haar*, a cold fog off the North Sea, was shrouding the fields and hedgerows in a blanket of grey when we set out on a tour of the farm. In the first field there were a dozen or so Shorthorn cows and roughly the same number of Aberdeen Angus, each with a calf. In the next, a sloping pasture with a scattering of tall oak trees, there was a larger herd of Shorthorns and a young bull. This was Jetstream. Arabella explained that he was a recent acquisition. She and Spencer had been looking for a bull, but there was no expectation that she would buy one when she went to the Great Yorkshire Show in July, accompanied by her two young children. However, she was greatly taken by Jetstream and agreed a price, after some haggling, with the owner. That evening she phoned Spencer with the news. 'He just gasped and said: "Jesus Christ, Ara!"' recalled Arabella, laughing at the memory. 'Well, that was our whole wedding budget gone in one go.'

'I suppose you could have got out of it,' I said.

'No. Absolutely not. We had shaken hands on it.'

In the beef business, once you have shaken hands on a deal – and you can even say 'let's shake hands' over the phone – the deal is done. 'If you reneged on a handshake, nobody would ever do business with you again,' said Spencer. 'It is an unwritten code of honour.'

'So what's so good about this bull?' I asked. We were standing about 5 yards away from Jetstream and Spencer suggested that we probably shouldn't get any closer.

'It's got tremendous length, a wonderful straight back, a great shape for his age and a lovely walk,' said Arabella.

Before I left, Arabella gave me a copy of the 2016 edition of *Red Cattle Genetics*. This listed nineteen Shorthorn bulls whose semen was available for artificial insemination, with the price per straw – the amount you need for each insemination – ranging from £6 to £40. Next to a picture of each bull was an Estimated Breeding Values chart providing an analysis of such traits as ease of calving, birth weight, milk yield, carcass weight after different periods of time, ribbed fat, eye muscle area and size of scrotum. The size of a bull's scrotum, incidentally, has a bearing on the fertility of its female calves. As a rule, the larger its balls, the more fertile its female progeny. I still can't get my head around this curious genetic trait. Jetstream's scrotum was measured for insurance purposes at 13 inches in diameter. Spending the wedding budget on Jetstream was not a punt in the dark, but a sound decision based on its genetic history and carefully calculated index of performance.

While researching *Batteries Not Included,* a report (you may remember from Chapter 6) about organic farming and animal welfare commissioned by the Soil Association, I visited a sheep farm in Snowdonia, a dairy farm in Sussex, the Gloucestershire pig farm that I mentioned in the last chapter, and poultry enterprises in Somerset, Lincolnshire and West Wales. For my research on beef cattle, I headed to the Scottish Borders and spent a day with Carey Coombs, who farms some 900 acres of meadow and rough grazing near Carnwath in Lanarkshire. He was one of the most interesting organic farmers I met, and the least didactic. He stressed that animal welfare depends on the skills of the men and women who look after the animals, as much as it does on the

Carey Coombs, organic beef farmer.

system itself, and that well-managed conventional farms will have better animal welfare than badly managed organic farms.

I rang Carey and asked if I could visit again. I was particularly keen to see him, not just because he is a well-known beef breeder, but because I knew he had interesting things to say on issues related to the sustainability of farming in the uplands and the economic challenges facing beef farmers. By the time I arrived at Weston Farm, at the southern end of the Pentland Hills, a misty sun was shining benignly over the high pastures. Over a cup of coffee, I asked him to remind me how he came to be here. It was an unusual story.

Carey was brought up in Reading and educated at Bangor University, where he studied marine biology. This was in the mid-1970s. He read lots of 'idealistic, back-to-the-land, Schumacher sort of stuff' – E. F. Schumacher was the author of *Small is Beautiful: Economics as if People Mattered* – and after university he and his girlfriend Penny headed up to the Borders in a gypsy caravan. He worked for the Forestry Commission, got married, had two daughters, rented a couple of fields and bought a few dairy cows. Then they moved to a larger farm in Argyll and began breeding Beef Shorthorns in a modest way. 'It was God's own country, but only on a good day,' recalled Carey. 'It rained for much of the time.'

In 1989, the Coombs took on the tenancy of Weston Farm, together with sixty Simmental-cross cattle which they inherited from the landowner. They decided to phase these out and focus on Beef Shorthorns. Carey, who had remarried since I last saw him, now runs 100 Shorthorns and their followers. He sells pedigree breeding stock and either finishes animals that are not sold for breeding or sells them on to other farmers. He is keen on performance recording, so all his animals are weighed at various times in their development and measurements are taken of muscle depth, testicle size and so forth. 'I'm trying to breed an animal of moderate size which thrives on grass, produces one calf a year, calves easily, is good on its feet and legs and has the right shape of head and colour,' he said.

When Carey first arrived at Weston Farm, he used some nitrogen fertiliser on the grassland and routinely vaccinated livestock against a range of diseases. However, he was already concerned about some of the things that were happening in modern farming. A knackerman had told him that ground-up material from slaughtered cattle was being used in animal feed. When he contacted some feed merchants in Edinburgh, only one of them would divulge what went into their feed. 'And then it all went pear-shaped when BSE broke out,' he recalled. An epidemic of bovine spongiform encephalopathy, or mad cow disease, infected over 180,000 cattle in the UK, and 4.4 million cattle had to be slaughtered as part of an eradication programme. An official enquiry concluded that the disease had been caused by feeding animal remains to ruminant livestock.

The BSE crisis encouraged the Coombs to become organic farmers. There was also another incentive for doing so. They found it galling that their traditional suckler beef cattle, reared outdoors on grass in one of the most natural forms of husbandry, were classified in the marketplace exactly the same as animals fattened indoor on grain and kept on hard slatted floors. They were not being rewarded for either their high standards of animal welfare or the quality of their beef. Going organic would provide them with a premium. Some organic farmers speak disparagingly of conventional, non-organic farming practices, but Carey is not like that. 'I see organic as an approach, a direction of travel,' he said. 'I don't like to think of things in black and white terms as being right or wrong.'

As we wandered around the cows, I asked whether he was working as hard now as he did when he first came here. 'I do just as much, I think, but everything takes me a bit longer,' he replied. His wife Hilary manages a bed-and-breakfast business at the farm, and his Dutch son-in-law Ed helps part-time, but he is still working long hours. Now that he was in his early sixties, he was thinking about the future, and trying to figure out how he and his wife could stay on at the farm without working so hard. This is one of the great challenges facing many tenant farmers. If you own a farm and you no longer have the strength, or the desire, to continue running it, you have several options: you can sell it, hand it down to your children, or get another farmer to manage it. If you are a tenant, you need to keep on paying the rent, and that means farming.

'I feel I am part of this place,' said Carey when he was cooking supper, a fine, spicy Spanish-style meatball stew. 'I've planted hedges and trees and created shelter belts. I have spent thirty years developing a herd of cattle. They depend on me and I depend

185

on them. That's how it is for us farmers. We manage the ecology of the land, we are part of the whole energy flow and life-cycle.' He said one of his pride and joys was a flower-rich meadow which he created out of an ordinary piece of old pasture. Now, thanks to his careful management, it is full of marsh orchid, yellow rattle and a host of other wildflowers.

Many of the environmental improvements on the farm had been financed by stewardship agreements and Carey had just applied to join a new five-year scheme which will encourage farmers in the area to create breeding habitats for wading birds such as curlews, lapwings, redshank and snipe. 'I'm very grateful for the stewardship schemes, but you shouldn't think it's money for old rope,' he said. 'It really does affect the way we farm and how much we can earn from livestock.'

This led us onto the subject of subsidies. It is impossible to overstate how important these have been since the Second World War. Much of the public money now spent on agriculture – either under the EU's Common Agricultural Policy (CAP) or by government bodies such as Natural England and Scottish National Heritage – has a social or environmental purpose. As we saw in an earlier chapter, most farmers in the Yorkshire Dales would not survive without the annual payments linked to the area they farm, and in many cases stewardship subsidies as well. However, this is a relatively recent development: originally, most subsidies were designed to stimulate production.

Prior to the early 1990s, EU farmers received a guaranteed price for their beef cattle which was significantly higher than

world market prices. This acted as an incentive to increase production. The CAP also guaranteed high prices for cereals, but
provided no support for oilseeds, such as soya beans, which are
used in animal feed. Trade policies were complementary: they
protected our livestock and cereal farmers with high tariffs, and
imposed low tariffs or none at all on oilseeds and feedstock. This
encouraged British farmers to adopt intensive systems of livestock
production based on the use of cheap imported feed.

This began to change in 1992, when guaranteed prices were
scaled-down in favour of area-based payments, although the
latter only applied to land that was producing arable crops.
However, in 2003 area-based payments were extended to
all types of agricultural land, including grassland, and this
provided an incentive for less intensive forms of livestock production, such as those practised by farmers like Carey. Indeed,
the most recent CAP reforms acknowledged that the loss of
grassland had become a serious problem and farmers now only
receive the full area-based payments if they are conserving their
grasslands.

When I visited the Scottish Borders, it was far from clear what
the government's approach was going to be when it came to nego-
tiating agricultural arrangements post-Brexit. Current levels of
subsidy were to be retained until 2020, but there was no guarantee
that these would continue beyond that date. Furthermore, several
politicians had raised the possibility that farmers might have to
survive without subsidies, and there was already talk of establishing free trade arrangements with countries such as New Zealand

and the United States, whose farmers can produce livestock and cereals at lower cost than ours.

When I asked James Playfair-Hannay if he thought he could survive if there were no subsidies to boost farm income, and no tariffs to protect farmers from foreign competition, he replied: 'Yes, I think we could – because of the way we farm. We have low stocking rates, free-draining soil and low rainfall and that means that we can keep our stock out all year round, and don't have the costs associated with bringing stock in for winter. That's the key for us: everything lives outside on grass, with a bit of home-grown cereals thrown in.' Farmers finishing beef cattle indoors on cereals might well find it more difficult to survive in a world without subsidies or tariffs.

There is a popular perception that landowners are a wealthy bunch of people and unduly privileged, whether through receipt of subsidies or the benevolent taxation arrangements on land. And it is true, many landowners are wealthy. But plenty of the landowners I met on my travels – this includes people like Stephen Ramsden in Nidderdale and James Playfair-Hannay – are obliged to work much harder than their forebears just to stay afloat. When James was growing up, there were twenty-four people working on the family estate; there are now just three, as well as himself. Furthermore, his wife Debbie works too, running a business that hires out high-class portable toilets for weddings and other events.

But at least landowners own their land, so even if they are property-rich and cash-poor, they are in a much better position than most tenant farmers, many of whom are now viewing the future with great trepidation. 'I couldn't have done what I have done without grant aid of one sort or another,' said Carey. 'If all subsidies are withdrawn after we leave the EU, an area like this would see massive change. Most of the tenant farmers on this estate wouldn't survive and you would find large areas being ranched – rather like New Zealand – by a much smaller number of people.'

I can recall in great detail a walk I took across the shallow valley to the south of Carey's farm and up into the hills beyond. It was a glorious end-of-summer day, warm enough to be in shirt sleeves, with swirls of rooks and the occasional raven drifting across the sky and buzzards circling over the high ground. Seen from a distance, the valley bottom looked like an over-peppered lettuce leaf, its green pastures liberally sprinkled with sheep and cattle. I suppose I saw about half a dozen people working on the land, including a man on a large tractor stacking round straw bales; another inspecting his cattle; a man on a quadbike with two sheepdogs. I am not sure how many farms I walked across or past: again, perhaps half a dozen.

No matter which paths I followed and where I looked, there was the architecture of a man-made landscape: the farmhouses and barns, stone walls and hedges, footpaths and tracks, ditches and drains – all of which require maintenance. This was a varied landscape, a mosaic of permanent pasture, grass leys, arable fields, commercial woodlands and – harder to spot – areas devoted to conservation, where farmers like Carey had adopted practices to encourage wildlife.

In a subsidy-free world, farming – or ranching – would be all about cutting costs and reducing labour. Over time, I imagine many of the hedges would be removed, though some might be left to grow ragged. There would be no point in spending time and money looking after the stone walls, so they would fall into disrepair, with barbed wire fences used as a way of controlling live-stock. Without subsidies, it would no longer make sense to retain species-rich flower meadows and boggy bits of marshland. They would be ploughed up, reseeded and fertilised to produce high-yielding grass monocultures for grazing or silage; or converted into arable land. So the wading birds, the lapwings and snipe and redshank, would lose their breeding habitat, and other wildlife would disappear too. Is this too bleak a picture? I would like to think so, but I fear it isn't. If we want these sort of landscapes to survive as they are today, farmers will have to be rewarded for looking after them.

8   Ploughing a New Furrow

In the early years of the twentieth century, there were over a million heavy horses working on the land and just a few dozen tractors. Half a century later, there were almost as many tractors as horses, with around a third of a million of each. By 1960 the number of horses left on our farms had fallen to 54,000 and soon there would be none at all. During the span of little more than a generation, horsepower had given way to oil power and mechanisation had paved the way for a dramatic increase in agricultural productivity.

I can still recall seeing, when I was very young, old-fashioned binders harvesting corn. The sheaves, laid out on the ground, would be stacked against one another by hand in batches of eight or ten. The stooks, as they were called, were left to dry in the sun before being taken to the granary, where the corn was threshed from the stalks. However, this was a relatively rare sight – in my memory I see it in sepia rather than colour, like an old photograph – and by the mid-1950s combine harvesters, which reaped, threshed and winnowed the crop all in one go, were already widely used across the lowlands.

As the years passed, the machinery used to plough, sow, fertilise and harvest our crops became progressively larger and more sophisticated. Grants, incentives and guaranteed prices encouraged mechanisation and the clearance of hitherto unproductive land. At the same time, new high-yielding crop varieties and the almost universal use of mineral fertilisers and other agrochemicals led to a spectacular increase in yields. Between 1885 and 1985 average wheat yields more than tripled, from 2 tonnes per hectare to over 6 tonnes; now they hover around 8 tonnes.

Inevitably, there was an environmental price to pay as marshlands were drained, hedges uprooted, ponds filled, woodlands cleared, heathlands ploughed up and wild places polluted with agrochemicals. One of the most devastating critiques of this brave new world of modern farming was provided by Marion Shoard in 1980. 'Although few people realise it,' she wrote in *The Theft of the Countryside*, 'the English landscape is under sentence of death. Indeed, the sentence has already been carried out. The executioner is not the industrialist or the property speculator, whose activities have touched only the fringes of our countryside. Instead it is the figure traditionally viewed as the custodian of the rural scene – the farmer.' The countryside, she wrote, was being turned into a vast, featureless expanse of prairie, given over to either growing cereals or a grass monoculture to support the intensive rearing of livestock. Unless something was done, it would be no more than a food factory by the early part of the twenty-first century.

Almost two decades later, Graham Harvey, agricultural adviser to the *Archers* radio programme and champion of small-scale, organic and low-input farming practices, wrote another much-lauded critique of the industry. In *The Killing of the Countryside*, Harvey claimed that British farming was in the process of being reinvented by a small group of business moguls who favoured a countryside devoted entirely to industrial-scale food production. 'The only space for wildlife in this new, utilitarian landscape is within a scattering of nature reserves or in ersatz habitats bought at high price,' he wrote. He laid the blame not just on a subsidy system that encouraged production regardless of the cost to the environment, and the multinational manufacturers of agrochemicals, but on farmers themselves; they, after all, were the ones who were destroying habitats, locking up livestock and drenching their fields with chemicals.

Flick through the pages of the latest *State of Nature* report, compiled by some fifty conservation agencies, and it would seem that nature is still suffering at the hands of farmers. According to the report, it is faring worse in the UK than in most other countries: a clumsily termed 'global biodiversity intactness indicator' ranks us 189th out of 216 countries. Between 1970 and 2013, 56 per cent of the 4,000 or so species sampled declined, and an index of species' status, based on abundance and occupancy data, fell by 16 per cent. The most significant 'driver of change' – in other words, cause of decline – was the intensive management of agricultural land, with the abandonment of mixed farming, the switch from spring to autumn sowing, the production of silage instead of hay, the increased use of pesticides and fertilisers, more intensive grazing regimes and the loss of marginal habitats being largely to blame.

Time, you might think, for another emotionally charged anti-farming polemic. Well, not from me. I am not denying that much of what Shoard and Harvey wrote was true at the time, although I saw no evidence of an industry dominated by a cabal of business moguls. And the *State of Nature* report certainly makes for depressing reading, although it is worth pointing out that when it comes to farmland bird populations – birds are a good indicator of the general state of wildlife as they have a wide habitat distribution – the most serious declines took place between the late 1970s and the early 1990s, when there was a period of rapid agricultural intensification. However, Shoard's grim vision of a countryside devoted to nothing other than intensive food production has not come to pass, and I don't think it ever will. Since the turn of the century, a growing number of farmers have begun not only to think, and worry, about the impact their activities are having on the natural world, but to recognise that it is possible to farm productively without destroying the biological life in the soil; that it is possible to get good crop yields and at the same time provide a habitat for wildlife. That is what this chapter, which focuses on arable farming, is really about.

The word arable comes from the Middle English *erable*, which relates to ploughing, an activity which also determined the way we measure land: acre comes from the Old English *aecer*, the area that could be ploughed in one day by two oxen using a wooden plough. Ploughing arable land also gave rise to a unit of distance still used in horse-racing, if not in agriculture, a furlong, or *furh lang* in Old English, being the length of a furrow – 220 yards – in an open field. In official statistics today, arable crops include cereals, oilseeds like rape and linseed, potatoes, sugar beet, peas that are harvested dry and field beans. In terms of the area they occupy, cereals are the most important arable crops.

The nature of arable farming has varied, both in terms of its extent and in the crops grown. During the last quarter of the nineteenth century, the availability of cheap imports from North America led to the collapse of cereal prices and over a million hectares of arable land was converted to grass. The opposite happened during the Second World War, when hostilities led to a dramatic fall in imports and increase in home-grown food production. Between 1939 and 1945 the arable area increased by 50 per cent. When most farm work was done by horses, oats (their staple diet) occupied an area of land which at times exceeded the area under wheat and barley. Now, there are 22 acres of wheat and barley for every acre of oats. For long periods of our history British farmers grew as much barley as wheat, and at times considerably more, but that changed during the 1980s as farmers took advantage of modern wheat's potential to produce higher yields. At the start of the decade, the area under barley was 60 per cent greater than the area under wheat. By the end it was 25 per cent less.

Wheat provides 20 per cent of the world's food calories and is the national staple of many countries, including our own, with wheat flour being used in the manufacture of scores of foodstuffs, from bread to biscuits, cakes to crumpets, pasta to pastries. It is also a major component of animal feed, as is barley, which provides much of the malt used in the brewing and distilling industries.

Cereal production generated some £2.9 billion a year for the UK economy in 2015 – the comparable figures for milk, beef and lamb were £3.7 billion, £2.7 billion and £1.1 billion – with exports of wheat accounting for about 15 per cent of the crop. However, my main concern here is not so much the economics of growing cereals, but the sustainability of cereal production and the impact it has on the ecological health of the countryside.

Everything begins and ends with the soil. 'The soil is the great connector of lives, the source and destination of all,' wrote Wendell Berry in *The Unsettling of America: Culture and Agriculture.* 'It is the healer and restorer and resurrector, by which disease passes into health, age into youth, death into life. Without proper care for it we can have no community, because without proper care for it we can have no life.' A similar view was expressed, less poetically but more succinctly, by President Franklin D. Roosevelt, reflecting on the environmental damage and human suffering caused in the 1930s by the American Dust Bowl, when a combination of drought and poor farming practices led to the loss of topsoil over an area of 100 million acres. 'A nation that destroys its soils destroys itself,' he said.

Here, in our dampish island on the East Atlantic seaboard, we have a mild climate, plentiful rainfall spread throughout the year, and deeper, more resilient soils than those of the American Plains. You may see eddies of topsoil whirling into the sky on windy days in parts of East Anglia, but we are not heading for a disaster on the scale of the Dust Bowl. However, there are some disturbing parallels with the American experience. In many parts of lowland England the soils have been impoverished by continuous cropping and the application of mineral fertilisers and other agrochemicals. This is now encouraging some landowners and farmers to change their ways. One of them is Hylton Murray-Philipson, whom you briefly met in the first chapter.

I first visited Blaston, the Leicestershire estate owned by Hylton, in the company of Dan Belcher, whose family has been grazing livestock there since 2012. By enlisting the help of the Belchers, and getting livestock onto the land, Hylton hoped to begin the process of restoring the farm's soils. Before the Belchers arrived at Blaston, there were about 100 acres of permanent pasture and 700 acres of wheat and oilseed rape. There had been no cattle on the land for forty years and the permanent pasture was in a poor state of health, having been grazed constantly by sheep for many decades. The soil in the fields used for arable crops had been much degraded. Partly as a result, the farm was suffering from an infestation of black grass, a weed which can dramatically reduce crop yields and profits.

On my second visit to Blaston I met John Lane, who used to manage the farm and now, in semi-retirement, still oversees much of the activity there, and Tom Heathcote, the land agent. These two men clearly had an excellent working relationship. John was a stockily built sixty-nine-year-old with a military bearing, roseate complexion, toothbrush moustache and sharp wit. He looked like

John Lane and Tom Heathcote at Blaston.

a part-time gamekeeper; in fact, he was a part-time butler. Tom, in contrast, was not much more than half John's age, tall and lean with the sort of hairstyle associated with a more arty way of life than managing country estates. Or, as John put it: 'Tom is not like a typical land agent – you know, tweeds, Audi, black Labradors in the boot. No, he's an excellent bloke.'

During the course of a blustery and cold April morning, John and Tom talked me through the various measures that were being undertaken to improve the soil. 'The plan is to increase the area under grass, reduce the area of arable crops and incorporate livestock in everything we do,' explained Tom. In the past, the rotation was short and repetitive: wheat, rape, wheat, rape, wheat, rape. Now, these combinable crops are being rotated with grass leys and crops which are helping to improve soil structure and increase fertility. The latter fall into two categories – companion crops and cover crops – and they have very different, though complementary, roles.

At Blaston, companion crops like clover and vetch are now being planted alongside oilseed rape. Their rapid root growth breaks down any compacted layers in the soil and creates channels that the roots of the main crop, rape, can then follow. Use of companion crops often reduces the need for insecticides, as insects are confused by a mixed crop. Clover and vetch, which tend to be killed off by the cold in late autumn, also release nitrogen into the soil, thus providing valuable nutrients for the growing crop during the winter months. Nature, in short, is the healthy surrogate for agrochemicals.

During the past few years, oil radish, berseem clover and black oats have been sown in the autumn at Blaston as cover crops. These help to increase organic matter, reduce soil erosion and improve soil structure. Beans and linseed are used as break crops to interrupt the repeated sowing of cereals and reduce disease and weeds. Just as importantly, the farm is moving away from deep ploughing to drilling directly into the seed bed created by the cover crops the previous winter. This not only reduces carbon losses, but the cost of cultivation, and keeps weed seeds on the surface, where they can be eliminated with glyphosate, more about which shortly. Some of the crops that have been introduced are poor earners compared to wheat and rape, but a more diverse crop rotation makes for a more sustainable enterprise in the long term. It also helps to reduce the bill for fertilisers and pesticides.

A year after my first visit to Blaston, I was invited back for an agronomy seminar. Those present included Tom and colleagues from the land agents Fisher German, a four-strong team of agronomists from Indigro, a company that has been providing the farm with technical advice over the past few years, and a soil scientist of international renown. Tom opened proceedings by giving a brief overview of what Blaston was hoping to achieve. 'We are assuming that after 2020 there will be zero subsidies,' he said. 'Our aim is to be financially viable, restore life in the soil, adopt practices which are good for wildlife, and significantly increase livestock on the farm, with a view to adding value and selling beef direct to the public. We are on a journey towards conservation agriculture.'

This was the first time during my travels in England that I had heard anyone use the term conservation agriculture, although it is widely practised in many parts of the tropics. Conservation agriculture has three main principles: minimal disturbance of the soil, keeping the soil covered throughout the year, and rotating crops. A few years ago, I visited peasant farming families who had adopted conservation agriculture in Zambia. As a result, they were getting 10 tonnes of maize per hectare, over six times the average yield for Africa and on a par with commercial farms in Europe. Conservation agriculture had not only helped them to restore their degraded soils but improve their incomes. They could now afford school fees, pay for better health care and buy the sort of consumer goods they had only dreamt of in the past. All of this was being achieved with family labour and virtually no machinery.

One of the jobs I enjoyed when I was a farm student, largely because of the time of year and the company of casual labourers, was hand-picking, or rogueing, wild oats. It was a task entirely lacking in skill, so we had plenty of time to chat as we walked in a line from one end of a wheat field to the other pulling up the tall weeds. In those days, you often used to see fields full of wild oats, but since my rogueing days they are no longer much of a problem as they can be eradicated with herbicides.

A much greater threat today, and a focus for discussions at Blaston, comes from black grass, whose dark seed heads you will frequently see waving above cereal crops in June and July. A native of Western Europe, black grass thrives on moist, heavy soils and now contaminates over half the cereal crop in the UK, frequently causing losses of 2 tonnes per hectare, representing a decline in yields of around 20 per cent. 'A single black grass plant might have forty to fifty stems, each producing up to 200 viable seeds,' explained agronomist Roger Davis. 'If you get ten plants in a square metre, that means there could be 100,000 seeds capable of producing black grass plants the following year.' One of Roger's tasks, both at Blaston and on other farms where his company Indigro works, is to provide advice on the best way of reducing black grass infestation.

Around 80 per cent of black grass seeds germinate in the autumn, which is why autumn-sown cereal crops are so vulnerable. One way of tackling the problem is to plant a cover crop, such as oil radish or forage rye, in the autumn. When the cover crops are destroyed in the spring, for example with the use of glyphosate – if you're a gardener, you will know it as Roundup – the black grass will be eliminated as well. The farmer can then plant a spring cereal crop using a direct drill to minimise soil disturbance and blackgrass germination.

'Spending £180 a hectare in the autumn on herbicides to tackle black grass is a major investment and is only well spent if it works,' said Roger. 'When it doesn't work, and we've now got a growing problem of black grass resistance to herbicides, farmers might have to destroy their entire crop, or accept much lower yields.' Yields of crops sown in spring will be less than those for autumn-sown crops, but this is compensated for by the lower costs involved. Sowing cereals in spring certainly makes environmental sense, as it precludes the use of large quantities of herbicides, fungicides and fertiliser. Introducing grass leys into the rotation, a key element of conservation agriculture, is another way of tackling black grass. Indeed, it was the abandonment of mixed farming over much of the countryside that created the conditions which have enabled black grass to flourish.

There is, however, a cloud on the horizon as far as farmers are concerned. The use of glyphosate, which is frequently used to destroy cover crops after they have done their soil-improving business, is now under threat. Although it has been approved for use by the European Food Safety Authority (EFSA) and numerous other regulatory organisations across the world, including the World Health Organization, the International Agency for

Research on Cancer (IARC) has concluded that it is 'probably carcinogenic to humans'. While EFSA looks at whether the levels of a product encountered by the people who use it are likely to cause cancer – this is a risk-based approach – IARC takes a hazard-based approach and looks at whether the products can cause cancer under any circumstances. IARC believes glyphosate probably can. But according to its system of classification, working as a hairdresser, working night shifts and drinking alcohol is just as likely to cause cancer as the use of glyphosate. German beer is said, disapprovingly, to contain glyphosate, but to reach the acute dose you would need to drink 1,760 pints a day. Despite all this, there are growing calls for a ban, backed by green politicians and activists. Having previously been licensed for use by farmers for fifteen years, the product has only been re-approved for another eighteen months, until the end of 2017. At the time of writing, we do not know whether it will be given further approval by the EU after then.

Does this matter? Yes, according to Neil Fuller, the soil fertility expert who came to talk to us about soils at Blaston. Before lunch – home-made shepherd's pie in the estate's shooting lodge – Neil took us on a brief tour of the fields to the north of the main farm buildings. As we were making our way from one field to another, I asked Neil, a tall and imposing figure with decades of experience advising on soils around the world, what effect a glyphosate ban would have on British farming. 'It would be absolutely massive,' he replied. 'Glyphosate enables farmers to use seriously beneficial approaches to improving soil health. Without it, it would be very difficult to do the sort of things they're doing here.' Farmers currently practising conservation agriculture would probably have to revert to deep ploughing, which can lead to soil erosion, the loss of moisture and higher fuel bills.

We began our brief tour in a field next to the home farm. This had always been permanent pasture. Neil plunged a spade into the ground and lifted up a clod of earth, before pronouncing that the soil was in fine condition, with good rooting, lots of earthworms and excellent soil structure. He became even more effusive when we reached the next field, which was called Stobo. Once again, he lifted a large sod of earth and began examining the soil profile with his fingers. 'There's a huge amount of earthworm activity, a sign of a healthy soil,' he said. 'You can see all these earthworm middens and galleries. The aeration is very good and the soil will warm up much faster in spring than soils with less organic matter. There's lots of biological activity and the crop will have a whale of a time.'

After thirty years of growing nothing but wheat and rape, Stobo was drilled with a species-rich grass-seed mix combined with white clover in 2011. For the next five years, it received no mineral fertilisers or agrochemicals, but was grazed by the Belcher's livestock. In autumn 2015 it was grazed hard to remove as much grass as possible and then given a light application of glyphosate to kill the grass, but not the clover. The field was then direct drilled, which means that seeds are drilled into unploughed soil, with winter oilseed rape, and the following season with winter wheat. The organic matter content of the soil increased from

2.4 per cent in 2011 to 4.4 per cent in 2015. This meant that the total amount of carbon contained within the soil had risen from 189 tonnes to 350 tonnes. Not only is the field now in much better shape for long-term food production, it is doing us a climatic favour. Conservation agriculture could well be one of the best ways of sequestering carbon and tackling global warming.

After my first experience at Blaston, in spring 2016, I wondered whether I had stumbled across an unusually enlightened enterprise, a bright star in an otherwise cloudy universe. At the time, I had no way of gauging whether the activities I had seen here were rare or commonplace, although it was obvious that Hylton Murray-Philipson was an unusual landowner. He had been brought up on the estate, but spent most of his working life – he is now in his fifties – making a living elsewhere. In his early twenties, he set up an office for Morgan Grenfell in Brazil, and subsequently worked as an investment banker in New York. He has always been a passionate environmentalist. He has been heavily involved in looking for cost-effective ways to save rainforests in the Amazon and worked with Amazonian tribes. He has farming interests in Australia, involving the restoration of degraded land, and he is a non-executive director of Agrivert, a company which converts waste food, sewage and household waste into electricity and a substance that is used as a replacement for mineral fertiliser on some 2,000 acres of arable land.

Hylton is precisely the sort of landowner you might expect to embrace concepts like conservation agriculture. But I now realise that he is not alone. A couple of months after my first visit to Blaston, I spent time at Cereals 2016, an event in the Cambridge countryside which attracted some 25,000 farmers, agronomists and professionals from the industry. Much of it was devoted to showcasing the latest generation of agricultural machinery; not just enormous and staggeringly expensive combine harvesters, crop sprayers and tractors, but drones that weighed no more than a bag of potatoes and other sophisticated hardware associated with 'precision' farming.

However, there was another side to Cereals 2016, one you wouldn't have seen a decade ago. There were numerous stands promoting the sort of technologies and practices now being used at Blaston, and in the main meeting room you could hear discussions that focused on how to restore degraded land and improve soil fertility. One of the more compelling speakers was – in his own words – a tight Yorkshireman, with the accent to go with it, who farmed in the Vale of York.

'I was destroying my soils,' began David Blacker bluntly. 'The organic matter was down to 2 to 3 per cent and I had serious problems of soil erosion and water run-off. We had tractors running around like ferrets on speed, which meant the soil was badly compacted, but there were other times when it was so wet we couldn't even get in the fields to cultivate them.' The solution lay in adopting much longer rotations, introducing spring cereals, using minimum tillage, adding compost and sowing cover crops and companion crops. In just a few years, these activities had helped him to increase the organic matter in the soil to around 4 per cent. 'It took two generations of abuse to get the soils to where they were in 2012, and it'll take some time to repair them – but we're getting there,' he said. Many others are getting there too.

In early February, with the countryside still in winter's tight grip, I headed to South Cambridgeshire, which had recently been named one of the best places to live in the UK in a national survey. It boasts some of the highest levels of employment in the country, the highest salaries and highest life expectancy. It also has a plentiful supply of good schools and public houses. The survey didn't take agriculture into account, but had it done so South Cambridgeshire would have excelled in this department too. Its fertile soils have traditionally produced fine crops of cereals, oilseed rape, sugar beet, field vegetables and salads. In recent years,

Harvested cereal fields near Seahouses in Northumberland.

they have also begun to show the ill effects of continuous cropping with wheat and rape.

To discuss this and other issues I spent an afternoon with Tim Breitmeyer. A tall man with a military stamp about him, he was deputy president of the Country Land and Business Association (CLA) at the time of my visit and would become president later in the year. The CLA, whose members own or manage around half of rural England and Wales, is one of the most influential organisations in the British countryside. Besides running the family farm, which covers around 1,600 acres, Tim also has contracts to farm another 3,000 acres of land and he owns a major sugar-beet harvesting business.

When Tim returned to the family farm in the gently rolling countryside near the village of Barton in 1996, after serving eighteen years in the Grenadier Guards, wheat was selling for around £120 a tonne. Within a few years, the price had halved. 'Falling cereal prices pushed East Anglian farmers down the route of reducing the costs of production and relying on modern chemistry and inorganic fertiliser to boost yields,' he explained. Farmers also adopted a short rotation of wheat following rape, as these were the most profitable crops at the time. The simplification of farming practices – in the past, most East Anglian farmers had included beans and peas in their rotation – led to declining soil fertility and the spread of black grass.

If you were to drive past the Breitmeyer farm, and look at the infrastructure and the machinery, the storage facilities and crop sprayers and harvesters, you would assume that this was an immensely profitable business. There's nothing in sight to suggest that farmers like Tim have poor years as well as good years, so it came as a surprise when he said: 'At the moment, we are busy fools, working very hard but not making a good return on our investment. I can concentrate on reducing our costs and growing good crops, but I have little control over many of the factors affecting our profitability. How much I make depends on the weather, geopolitical events and the currency markets – all largely beyond my control.'

Between 2008 and 2013, wheat prices in the UK rose from around £80 a tonne to £180 a tonne, providing cereal farmers with excellent profits. Then in 2015 prices plummeted down to £100 a tonne, reflecting an increase in supply and fall in demand on the world market. For a business like Tim's, this represented a significant reduction in profits. While fine weather might boost yields, poor weather can do the opposite. For example, October 2013 was very wet in eastern England and this meant that some 300 acres of winter wheat on Tim's farm were sown too late to thrive. As a result, he got just 7 tonnes a hectare, rather than 10 tonnes.

Declining soil fertility and the threat posed by black grass have encouraged Tim to change his farming practices. 'If you'd come here a few years ago I would have been able to tell you exactly what crops would be sown in which fields for the next five or six years,' he said. Not any more. The farm now has longer and more complicated rotations, which include cover crops as well as peas and beans. He has also reduced his use of mineral fertilisers and buys in large quantities of duck manure and treated sewage, which help to increase the organic matter in the soil. 'We are doing all this to improve soil health and it's working,' he said with satisfaction. 'We are getting good yields and spending less on chemicals.' Tim didn't strike me as being ideologically 'green' like Hylton Murray-Philipson; he had simply come to the conclusion that the long-term profitability of his business depended on improving soil health.

As we made a tour of Tim's farm we talked about many things besides the condition of his soil and his cropping plans for the future. He is a keen advocate of diversification, and believes farmers should be encouraged to supplement their farming income by establishing other enterprises. He was speaking from experience. He took me to see an old piggery that he had converted into offices which he rents to small businesses and a nursery school, and he was the first farmer in Cambridgeshire to establish a large solar power unit on the roofs of some of his farm buildings.

Before I left I asked him how he felt about the vote to leave the EU. 'Brexit could be very good, if it helps to get farmers off income support and become more efficient,' he said. 'But in the short term it could be a rude awakening, especially if we don't get a decent trade deal with the EU.' British farmers were currently receiving over £3 billion in direct payments from the EU and the Treasury had agreed to ring fence these payments until 2020. 'Whatever happens after that, there is a danger there will be less money available to farmers than they currently get, whatever promises were made before the referendum,' said Tim. He believes that there is a good case to be made for providing direct payments to farmers for non-market public goods, which could include the provision of clean water, sequestering carbon and encouraging biodiversity. We already have a template for this in the form of stewardship agreements, the best of which are repairing the damage done to nature by years of intensive farming practices.

Prior to reforms of the Common Agricultural Policy (CAP) in the 1990s, most agricultural subsidies encouraged farmers – the pig, poultry and horticulture sectors excepted – to increase productivity: to produce more milk, beef and lamb, to grow more wheat, barley and sugar beet. However, there was a serious downside to these subsidies. They led to the creation of surpluses, such as the EU's infamous milk lakes and butter mountains, some of which were dumped on the world market, lowering prices for producers elsewhere. They also led to widespread environmental damage, such as overgrazing in the uplands and the destruction of wildlife habitats in the lowlands.

Subsidies are no longer linked to food production and the lion's share now goes to the Basic Payment Scheme. Most of the rest, some £600 million a year, is allocated to agri-environment schemes. Many of the farmers you met in the previous chapters have benefited from these. The sheep farmers in Nidderdale, for example, were enrolled in Higher Level Stewardship schemes which have encouraged them to reduce livestock numbers and restore old-fashioned hay meadows. This is leading to an increase in the populations of wading birds such as curlews and lapwings. In Devon, the Hunt family had been paid to restore scrubland habitat on the scarp above their cider press; this has helped to increase the population of cirl buntings. At Elveden Estate, in Norfolk, a Higher Level Stewardship scheme is paying for the restoration of wildlife-rich heathland, grassland and field margins on some 5,000 acres of land, benefiting a wide range of species, including the rare stone curlew.

When I began taking a professional interest in environmental issues in the 1970s, it was an article of faith among conservationists that the best place to conserve wildlife was in nature reserves, which were viewed as islands of biodiversity in a sterile sea of farmland. It was almost as though the agricultural landscape had been written off. However, there were some dissenting voices, the most notable being Dr Dick Potts, whom I was fortunate enough to meet in 2015 when I was writing a book about wildlife management.

Dick picked me up from Arundel station, in Sussex, on a glorious morning in early July. 'Eddie wants to see you,' he said as I climbed into his car, 'and he might even take a look around the farm with us.' We found the eighteenth Duke of Norfolk, Edward Fitzalan-Howard, an easy-going and convivial character, in the estate office. He recalled how Dick had first come to see him in 2002, shortly after he had retired as director general of the Game Conservancy Trust, now the Game and Wildlife Conservation Trust (GWCT). 'He told me that if we didn't act now, the grey partridge would soon become extinct on the South Downs,' he

Stewardship schemes have helped to transform parts of the South Downs.

recalled. 'I thought: as a shooting man and as the owner of part of this area, if I can't do something, then who can?'

The son of a Yorkshire farmer, Dick first came to the South Downs in 1968 to explore why grey partridge numbers had been declining. He discovered that the use of herbicides by arable farmers had led to the loss of hundreds of species of plants and insects which grey partridge and other birds had thrived on for thousands of years. The lack of suitable nesting cover and heavy predation by foxes and crows simply made matters worse. Dick established what has now become one of the longest running studies of farmland ecology in the world, and from 2002 almost up to his death in 2017 he acted as a consultant ecologist to the Norfolk Estate, which has undergone a remarkable transformation, thanks in part to his influence.

The Norfolk Estate has replaced intensive prairie-style arable farming with a more traditional system using smaller fields and carefully planned crop rotations. Farming is no less intensive, in terms of the use of fertilisers and other agrochemicals, than it was in the past in the areas where crops are grown, but around 10 per cent of the land that used to be under crops is now devoted to wildlife. Conservation headlands and beetle banks (grass-covered mounds bisecting arable fields) now support a rich variety of herbs and insects. Over 8 miles of hedgerows, which provide cover for nesting partridges, have also been planted on the Norfolk Estate. These are particularly important as there has been a significant increase in the number of birds of prey during recent years, including peregrine, hobby, sparrowhawk and barn owl.

In 2003, there were just three pairs of grey partridge on the estate and they produced eight chicks between them. By 2014, 292 pairs nested, producing on average 8.2 chicks each. 'The management here is like a three-legged stool,' said Dick, 'and without any of the three legs, the stool would collapse.' The key elements are a plentiful supply of insect food, good nesting cover and effective predator control. It is not just grey partridge which have benefited. The density of corn buntings is now ten times greater here than in other parts of the South Downs. Skylark numbers on the estate tripled between 2007 and 2011 and the lapwing population more than doubled during the same period.

This story illustrates how carefully targeted stewardship funds – the Norfolk Estate benefited from a ten-year Higher Level Stewardship agreement with Natural England – can transform relatively sterile areas of arable land into a landscape that supports both crop production and wildlife. 'What we're trying to do is find a middle way where we can reverse the decline in biodiversity, but still help feed the world and pay the wages,' said the Duke. 'The stewardship scheme provides compensation for the farming income we forego by dedicating land to conservation headlands and other non-productive areas.' While the Basic Payment Scheme simply rewards farmers for farming, with a few strings attached, stewardship schemes such as this are paying farmers for services rendered. This could be the shape of things to come once the current subsidy system comes to an end in 2020.

It was one of those all's-right-with-the-world days in early April, with a hazy sun shining down on the chalk ridge and skylarks trilling above our heads. From the northern fringe of the South Downs National Park we had fine views across the undulating landscape, with its mosaic of orchards and pasture, ancient woodlands and arable land. The apple and cherry was in white blossom, and so was the blackthorn in the hedges below us. At our feet were great drifts of yellow cowslips; soon, orchids and other downland flowers would appear as the days lengthened.

I had come to see William Wolmer, managing director of the Blackmoor Estate, to hear about a new venture in conservation, known as farmer clusters. The story really began here, in these meadows, he explained. Some nine years ago, the estate

William Wolmer in a field of cowslips.

entered into a Higher Level Stewardship agreement with Natural England. One of the projects involved the creation of a wildlife corridor between Selborne Common, whose beech woods and flower meadows inspired Gilbert White, the eighteenth-century parson-naturalist and author of *The Natural History and Antiquities of Selborne*, and Noar Hill, a nature reserve managed by the Hampshire and Isle of Wight Wildlife Trust.

'We came up with this idea of turning two arable fields into chalk grassland, and Natural England provided advice on how we could do this,' recalled William. One of the aims was to establish plants that would attract two species of butterfly which had become increasingly rare during recent decades. At first, the estate used an off-the-shelf seed mix. This didn't work particularly well, so a crop of green hay was taken from Selborne Common and spread on the fields. In one of the fields the cowslips now grow in such profusion that it's hard to avoid stepping on them. These are the food plant for Duke of Burgundy butterflies. The estate also planted blackthorn hedges which provide food for the larvae of the brown hairstreak.

'It's all about making connections between places of conservation value, such as nature reserves, and looking at conservation at the landscape scale,' said William, who is the fifth generation of his family – the first being the first Earl of Selborne – to farm here. It was the third Earl who established the estate's top fruit business, which supplies Tesco and many other supermarkets with apples, pears and cherries. The orchards cover just under 200 acres and a further 800 acres or so are devoted to arable crops. There are also large areas of ancient woodland as well as chalk grassland, acid grassland and heathland, reflecting the geological diversity of the area.

In 2013, William received a phone call from Teresa Dent, chief executive of GWCT. 'She told me about her idea of setting up farmer clusters, and asked me if I was interested,' he recalled. In fact, he was one of six farmers contacted by Teresa. Five of them immediately said they were keen to get involved. The idea was simple: farmers who were well respected in the local area and had good green credentials would encourage their neighbours to work together to achieve environmental goals that they would set themselves.

William's enthusiasm for the idea owed much to his previous experience in the field of conservation. After he left university, he worked as a researcher in various countries in Africa and for his PhD thesis he wrote about conservation and development in the Zimbabwean lowveldt. 'In Africa, there is a long history of "farmer first" and "bottom-up" development,' he reflected, 'whereas the post-Second World War consensus here was based on the idea that experts always know best.' Farmer clusters would put farmers, rather than experts, at the centre of decision-making.

After William had spoken to Teresa Dent, he began phoning round his neighbours, all of whom agreed to join the new farmer cluster, now known as the Selborne Landscape Partnership. At the time of my visit, members of the cluster included fourteen farmers and three other landowners, the National Trust, the county wildlife trust and the Gilbert White House Museum, who between them own and manage some 10,000 acres of land. Most of the productive land is down to arable crops, but there are also significant areas of pasture grazed by beef and sheep, as well as William's orchards.

'Many of the farmers I spoke to were already involved in conservation schemes, and I suggested that if we worked together the whole would be greater than the sum of the parts,' explained William. Over the next eighteen months or so, members of the cluster met regularly at Blackmoor village hall. First, they shared information, about topics such as fly tipping and crime as well as conservation. Then they drew up maps of their farms to identify the main areas of conservation interest and assess what they could do in future, with a view to creating corridors of wildlife habitat connecting the best sites for nature conservation. 'It's been a very practical exercise,' says William, 'and we are only doing things that we believe we can achieve.' The main focus is on enhancing and linking species-rich woodlands and grasslands, creating new habitats for insect pollinators and farmland birds, and improving soil management to enhance water quality.

When Teresa Dent came up with the idea of farmer clusters, she believed there was just one question the lead farmers should ask their neighbours: what sort of wildlife do you want? Everything else would flow from that. Although the farmers make all the key decisions, they benefit from technical advice from outsiders. In the case of the Selborne Landscape Partnership, one of some fifty farmer clusters that had been created by mid-2017, the GWCT has provided training on the identification of farmland birds and on how to find the nests of harvest mice, which were first recognised as a separate species by Gilbert White in 1767. The harvest mouse was thought to be almost extinct, with just one nest recorded in the area since 1990. Following the training exercise, over 470 nests have been identified, so the species is thriving. William and his neighbours are in the process of drawing up a list of other target species for conservation. This will probably include tree pipits, wood larks, woodcock and barn owls.

The dominant theme in the two critiques of modern farming that I mentioned at the beginning of this chapter was one of Paradise Lost. I am not suggesting that farmer clusters or environmental stewardship programmes have reversed all the damage done by decades of intensive farming, but I think there is plenty of room for optimism.

Creating healthy landscapes is about far more than restoring lost habitats and wildlife populations. It is also about creating more sustainable farming systems. This brings me back to where I began, in the lovely rolling countryside of south-east Leicestershire. By integrating livestock with arable cropping, the Blaston estate has brought life back to its degraded soils and created a system of farming which, in principle, could last forever. I hope we see many other arable farms adopt similar measures. There is, as the old proverb says, nothing new under the sun. This is all about going back to the future, to mixed farming systems which take advantage of the technological cleverness of our own times.

Epilogue

M ost of the farming families I encountered during my travels were not only surviving, but in many cases thriving, and facing the future with a degree of confidence, despite all the uncertainty created by low commodity prices and our future trading arrangements with the European Union, which currently accounts for 70 per cent of our food exports. Most were entrepreneurial, keen to innovate and eager to adapt to changing circumstances. However, there is another world which I have scarcely mentioned, of farmers struggling to survive, of real poverty and sometimes desperation.

During the agricultural depression of the 1930s, five farming families from Lancashire headed south to the Cotswolds in the hope of making a better life. Among them was four-year-old Malcolm Whitaker, who has lived in the village of Syde ever since. 'When I was a young man,' he recalled, 'there were twenty-five farms in the village. Now there are just four. In those days, everyone helped each other, but the old farming communities have almost ceased to exist now.' To fill the vacuum, Malcolm recruited a small group of farmers and farmers' wives who were willing to take phone calls from, and visit, farmers in distress. Gloucestershire Farming Friends is now part of the Farm

Community Network, a Christian organisation which has some 400 volunteers in England and Wales.

I met Malcolm at the farmhouse of Sue Pullen, the current chair of Gloucestershire Farming Friends and a beef farmer near the village of Churchdown. 'There is a high rate of suicide among farmers, but dealing with suicide is just a very small part of what we do,' explained Sue. 'During the last three months, we have probably had more phone calls than we've had in the past five years.' The majority were from farmers who had yet to receive the money they expected under the Basic Payment Scheme, largely because of the bureaucratic sloth of the government's Rural Payments Agency. As a result, banks were refusing to extend overdrafts and farmers were unable to pay their bills.

Since Gloucestershire Farming Friends was set up in 1991, there have been several major crises. 'The impact of foot-and-mouth disease was catastrophic, especially when whole dairy herds, including the calves and cows that were near full time, had to be shot,' recalled Malcolm. 'So we were always there, at the end of a phone for farmers who wanted advice, or just needed to talk. It was the same when we had the floods in 2007.' It tends to be older farmers with smaller enterprises who are most affected by financial problems, but loneliness is a serious issue too, as farming has become a very solitary affair. When I was young, there was plenty of bustle and manual industry in the countryside, with farm labourers working together, forking hay or straw, picking potatoes, trimming hedges and rogueing weeds. A fair field full of folk is now a distant memory in many parts of the country.

As old as the hills: Swaledale sheep looking down Swaledale.

If this had been a bigger book, it would have had more about the vicissitudes of farming life. I also wish I had had more space and time to describe many of the people I encountered, some by chance, some by design, who haven't featured in *Land of Plenty*. Such as the farmer who railed against the work ethic of the indigenous youth – 'they've all got degrees in effing social media studies' – yet resented the presence of so many Eastern Europeans, even though he admitted they were hard workers; the young lawyer who had given up his career in London and now produces high-class goat's cheese in Somerset; the Yorkshire dairy farmer who has to keep selling off bits of land just to pay the overdraft; the elderly farmer in the Pennines who insisted that his farming neighbours were getting far too much from the state for doing far too little; the organic farmers in Devon who had fallen on hard times and been forced to sell their truck and pull the back seats out of their saloon car to transport produce to the market; and the farmer near Glastonbury who makes the best Cheddar I have ever tasted, and who told me: 'I don't approve of cows being indoors all year round. It's wonderful to see them galloping about when you put them out in the spring.'

'The old order changeth, yielding place to the new', wrote Alfred Lord Tennyson. This could be a description of agriculture through the ages. Over the coming years, we can expect to see profound technological changes that will affect the way in which food is produced and processed. The use of robots, drones and unmanned vehicles, such as driverless tractors, will become more commonplace. Practices such as hydroponics, a cross between fish farming and growing crops without soil, could revolutionise the production of salads and vegetables. There will be further advances in plant and animal genetics and changes in the way we feed and raise livestock.

It is just over a year since I set off on my travels. The dawn chorus has reached its musical zenith; the hedgerows, festooned with hawthorn blossom, hum with insects; the woods reek of wild garlic; winter-born lambs are now half grown and the fields are full of cattle grazing on fresh grass. It is the season of renewal and a time for optimism for those of us who don't make a living from the land. If you do, you may be viewing the coming years with a degree of trepidation, not least because farmers' fortunes will depend on political decisions that they can do little to influence.

When I began researching *Land of Plenty*, the polls suggested that we would stay in the EU. Some of the politicians proposing that we should leave made much of the fact that once we were outside the EU we would be able to embrace free and unfettered trade and usher in an era of cheaper food. They either have a blinkered sense of history or are unconcerned about the fate of the farming world. In the 1870s, free trade led to the virtual collapse of the agricultural industry in Britain as our farmers were unable to compete with competition from America and elsewhere. 'The men of theory failed to perceive that agriculture is not merely one industry among many but is a way of life, unique and irreplaceable in its human and spiritual values,' wrote the historian G. M. Trevelyan. I believe this still remains the case.

Many farmers recognise that they will need to diversify their sources of income, and turn their hands to other activities besides farming, if they are to survive in a world with less state support. Nevertheless, food production will always be their main *raison d'être*. In 1995, British farmers provided close to 80 per cent of the food we consumed; this figure has now fallen to around 60 per cent. Meanwhile, our population continues to rise. There were just over 65 million people in the UK in 2015; the population is expected to reach 70 million by 2027. Instead of increasing food imports to meet growing demand, we should be encouraging and supporting more home production. We are fortunate to have the soils and climate which make this possible. We also have standards of food safety and animal welfare, as well as environmental safe-guards, that are among the highest in the world and far superior to those of many of our competitors. If you care about such matters, all the more reason why you should buy food which is grown and reared here. Living in a land of plenty should be a reality, not just an aspiration, and we consumers have as much of a part to play as our farmers.

Acknowledgements

A great many people provided me with help and information while I was researching and writing *Land of Plenty*. Some of the families I met gave me a bed for the night or a place to park my motorhome; many invited me into their homes for lunch or dinner. Indeed, I met with great kindness and generosity wherever I went and I would like to thank everyone who features in this book, and those I met but haven't written about, for sharing their knowledge and time.

In Leicestershire, Mike and Heather Belcher, Dan Belcher, Tom Belcher, Pete Osborne, Paul McDonald and Jane Riley provided me with an insight into the world of family farming. Among those who helped to shape my understanding of the dairy industry were Stephen and Claire Bland, Tom and Sue Flowers, Adam Sills, James Griffiths, Brian Hesford, Simon Weaver, Will Atkinson and Joan Turner. For the chapter on sheep I focused on Nidderdale in Yorkshire. Stephen Ramsden spent a considerable amount of time talking to me about upland landownership and introduced me to many of his tenant farmers, including Andrew Hattan, Alan Firth, Jack Challis, Mark Ewbank and Spencer Ewbank, all of whom were generous with their time and knowledge. I am also very grateful to John Rayner,

John Stevenson, David Verity and the many sheep farmers I met during my travels around country fairs.

The following individuals provided fascinating insights into the world of fruit production: Roger Hunt, Richard Hunt, Annette Hunt, Martin Thatcher, Penny Adair, Chris Newenham, Kevin Townsend and Veso Ivanov. Research for the chapter on vegetables took me from the Vale of York to East Anglia, and in particular I would like to thank Tony and Linda Beckett, Andrew Etty, Alan and Beryl Sykes, Patrick Sykes, Andrew Francis, Emma Clements, Andrew Burgess and Joe Rolfe. My understanding of the pig industry owes a great deal to the openness and honesty of Rob Shepherd, Peter Batty, Andrew Freemantle and Grant Morrow. I am particularly grateful to Gloucestershire vet Roger Blowey, who I visited on two occasions, first to discuss animal welfare and dairy cows, and later to talk about pig farming. In terms of the welfare of farmers, and the problems many face, I learned much from Malcolm and Di Whitaker and Sue Pullen of Gloucestershire Farming Friends.

James and Debbie Playfair-Hannay, Carey Coombs, Spencer McCreery and Arabella Harvey all entertained me and talked to me at great length about native-breed beef farming. Among those who helped me gain a better understanding of arable farming and conservation were Hylton Murray-Philipson, John Lane, Tom Heathcote, Roger Davis, Neil Fuller, Tim Breitmayer, Andrew Robertson and William Wolmer. Caroline Drummond, chief executive of Linking Environment and Farming (LEAF), introduced me to several of the farmers who feature in this book. I am also very grateful for the help I received from Teresa Dent and Andrew Gilruth of the Game and Wildlife Conservation Trust, the late Dick Potts, the Duke of Norfolk, Julia Aglionby, Mark Gorton, Ritchie Riggs, Joanne Rodger, Simon and Sarah Singlehurst, Christopher Price, Richard D. North, and George Dunn and Stephen Wyrill of the Tenant Farmers Association.

I owe a special debt of gratitude to my old farming friends in Yorkshire, and particularly to Harry Marwood and to two generations of the Keeble family. Mike Keeble, a famous character in the world of farming in the north of England, took me on as a green-behind-the-ears farm student in the early 1970s. Shortly before he died in the autumn of 2015, he suggested we drive up Wensleydale as he wanted to discuss, among other things, the impact of modern dairy farming on upland farming. He was that sort of a person: restlessly examining present practices and exploring how to improve them in the future. Although he doesn't feature in *Land of Plenty*, he and his wife Peta did much to shape my own views on the business of making a living in the countryside.

My final thanks go to the staff at Elliott & Thompson, who commissioned this book and provided continuous support throughout its writing and research. I am particularly indebted to Jennie Condell, publisher at Elliott & Thompson, for her insightful advice and criticism. This would have been a much poorer book without her help.

Index